Understanding the Dynamics of Nuclear Power and the Reduction of CO_2 Emissions

Hassan Qudrat-Ullah

Understanding the Dynamics of Nuclear Power and the Reduction of CO_2 Emissions

A System Dynamics Approach

Springer

Hassan Qudrat-Ullah
York University
Toronto, ON, Canada

ISBN 978-3-031-04340-6 ISBN 978-3-031-04341-3 (eBook)
https://doi.org/10.1007/978-3-031-04341-3

This Springer imprint is published by the registered company Springer Nature Switzerland AG
The registered company address is: Gewerbestrasse 11, 6330 Cham, Switzerland

To
Tahira Qudrat,
Who selflessly and with love helps me in most of the things I do daily. I am amazed by her bearing with and walking with genuine care for me, a demanding childlike person, for a happy 31+ years.

Preface

Energy plays a fundamental role in the well-being, prosperity, and improved living conditions of all of us. Consistent with the United Nations' SDGs especially Number 7: "ensuring the reliable, affordable and cleaner supply of energy for all", energy policy planners and decision makers of any region, state, or country aspire to have a common goal–adequate, reliable, cleaner, and affordable supply of energy. Pakistan, a developing country with increasing energy demand, faces challenges and opportunities in its energy sector. On the one hand, the country is facing severe issues in energy supply: (i) severe load-shedding and bailouts are becoming a norm in the electricity supply sector, (ii) affordability is an issue (due to relatively higher prices), and (iii) electricity-related emissions are increasing at an alarming rate. On the other, the country has a successful nuclear power experience with an abundance of available indigenous nuclear power fuel and fuel-manufacturing capability. Pakistan has also signed and ratified the Paris Agreement (PA) in 2016 which requires a reduction of GHG emissions from their peak level by 2030. To achieve a reduction in GHG emissions, Pakistan has submitted a report, Pakistan's Intended Nationally Determined Contribution (Pak-INDC) in 2016. The Pak-INDC presents principles and actions that are underway for reducing GHG emissions. It also describes mitigation and adaptation measures already being implemented in Pakistan. However, given the continuing nature of energy crises in Pakistan and the predominant nature of the fossil-based generation of electricity, it is necessary to assess the potential environmental (e.g., CO_2 emissions) impact of ongoing initiates and policies regarding energy supply in Pakistan. A modeling-based analysis, where underlying assumptions and mechanisms in the Pak-INDC are explicitly accounted for, allows us to verify the role of current mechanisms and policies in reducing the emissions in the frame of the PA. The general question, what will be the impact of these programs on the pursuit of reducing environmental emissions, is thoroughly addressed in this book.

Although our focus and preference for the main dynamic modeling-based analysis will be in the context of Pakistan, at times, however, for comparative analysis, India and Saudi Arabia's cases will be evaluated and discussed as well in this book. After China, India is the second-most populous country in the world with increasing energy demand. Saudi Arabia is the world's largest producer of oil and predominantly operates with fossils-based electricity generation. Besides the geopolitical importance of these countries, all of them have ambitious nuclear power plans as a possible strategy to combat electricity-related emissions. Another commonality among these three countries is the "dominance" of thermal power in their electricity supply mix. Therefore, the term "the PIS region", in this book, will refer to these three countries: Pakistan, India, and Saudi Arabia.

The Objective of This Book

The overall objective of this book is "to enhance systematically our understanding of and gain insights into, the general process by which the role of renewable and nuclear power in reducing CO_2 emissions can explain the making of a low carbon economy and a systematic achievement of PA's targets about the reduction of CO_2 emissions for a country or region". By identifying and critically examining the possible energy transition pathways available for the policymakers (including the PIS region) with the help of a dynamic modeling-based scenario approach, we attempt to contribute to the ongoing larger debate about climate change and low-carbon economic regimes.

The Approach of This Book

To accomplish the objective of this book, we apply the system dynamics modeling approach whereby a dynamic model, capable of representing the important interactions among various sectors of the electricity supply sector of Pakistan, supports our scientifically rigorous analysis. This book draws on Pakistan's case data. A systematic comparative overview of the PIS region (i.e., Pakistan, India, and Saudi Arabia) in terms of its power sector emissions, electricity supply mix, CO_2 reducing policy initiative and mechanisms, and nuclear power plants is also carried out. This design consideration in a policy design requires the policymakers to apply a systematic and holistic decision-making approach where the achievement of reliable, affordable, and cleaner electricity supply is the goal of their energy policies. This framework is generic enough to apply to the energy policy assessment and design exercise for any country in the PIS region and beyond.

Outline of Book

This book has eight chapters dealing with three distinct but related themes of "Energy-related CO_2 Emissions Reduction:"

1. The introduction and review of energy policies, mechanisms, and CO_2 emissions, understanding of the nature of CO_2 emissions reduction task, and analysis of energy policy instruments for the promotion of renewable and clean including nuclear energy (Chaps. 1, 2, and 3).
2. Role of Simulation and Modeling in service of energy systems, development of a dynamic simulation model and its validation to be used for energy policy design and assessment purposes (Chaps. 4, 5, and 6).
3. The development and evaluation of various CO emissions reduction scenarios with the dominant role of nuclear power and renewable energy generation technologies. Demonstration of the application of dynamic modeling-based innovation solution models together with the exploration of the avenues for future research (Chaps. 7 and 8).

Use and Users of Book

This book provides the reader with a comprehensive understanding of how to reduce CO_2 emissions with the dominant share of nuclear power in the supply mix of a country to meet desired targets such as in the Paris Agreement. For energy policymakers and practitioners, this book provides a set of rigorous dynamic modeling-based insights about how to accomplish the desired reduction of CO_2 emissions through a systematic energy policy design. Teachers and instructors in the post-secondary education sector can adopt this book as a textbook for a course on Policy Modeling or Energy-Environment Modeling. For researchers and students, this book provides probably the most systematic study of the role of nuclear energy in electricity-related CO emissions. The existing molars especially in the system dynamics field can avail the material on model validation (i.e., Chap. 6), which demonstrates how to apply both structural and behavioral validity tests to increase the validity and appeal of dynamic models.

Toronto, Canada
March 2022

Hassan Qudrat-Ullah

Acknowledgments

First of all, I thank Almighty Allah (SUT) for granting me the faculties and energies to embark on this journey. For any project you undertake, you need a constant source of inspiration, encouragement, and motivation. I am fortunate to have my mother, Fazeelat Begum, as the very source of all of it: at 72+ years of age, her concerns and sacrifices for the well-being of "others" provide me with huge energy and shine my way. She travels across the Atlantic to provide selfless and special moral support to all of Canada. Content-wise, this brief book draws heavily on my research in and practice of "energy policy modeling" that I started in1996–1997 at the University of Bergen, Norway, during my master's program; I am grateful to Prof. Paal Davidsen for introducing me to this research area, specifically for teaching system dynamics methodology.

I would like to thank my colleague from Springer: Niko Hisako, for their encouragement and support throughout this process. Ms. Jayanthi Krishnamoorthi (Mrs.), Project Coordinator—Total Service Books Production is appreciated for her professional support and timely production of the camera-ready copy for this book. I am thankful to Anam Qudrat and Ali Qudrat, who provided professional editorial support for this manuscript. The other people, who I especially want to thank for their prayers and good wishes for me throughout my academic and professional life include my sister, Zahida, and brothers Naheem, Saleem, Naveed, and Wasim. Tahir's support in my several mundane tasks is continuous. Last but unparalleled prayers and support of Saira Bano and Allah Ditta are worthy of mention here. Finally, Col. Tariq (late), Dr. Ashiq Javaid, Dr. Tahir M. Chaudhary, Z. A. Khan, Shahid Hussain, Khurram Mughal, Dr. Haider Madani, Shamim Abdullah, Alamgir Khan, Dr. Saleh M. Mustafa, Dr. Mustafa Karakul, and Dr. Aymen Kayal are always be remembered for their support and prayers for me, especially during the very challenging time of my early (student) life and now when I am getting old☺.

Contents

Chapter 1
Energy Policies, Mechanisms, and CO_2 Emissions

Check for updates

1.1 Introduction

Energy embodies a necessary and sufficient condition for almost all aspects of modern societies. Energy resources and the use of electricity lie at the heart of the sustainable development and prosperity of any region or country. Energy policymakers face, what can be termed as a classic "development-cleaner development" dilemma of the twenty-first century. While several developing nations, especially those with the domestic availability of fossil-based energy (e.g. coal, oil, gas), are consuming these resources for the prosperity and growth of their countries, the environmental and health hazards of emissions require more use of cleaner energy in their supply mix. Fossil fuel-based energy use also gives rise to other types of pollution and damage (Manisalidis et al., 2020). These can include but are not limited to land degradation, water pollution, ocean acidification, sea-level rise, emissions other than carbon dioxide (CO_2), floods, and droughts (Ali et al., 2020; NRDC, 2021). In addition, there is an increasing urgency and drive by the UN, through SDGs[1] (e.g., through SDG No. 7: clean and affordable energy for all people), for cleaner energy for all (Qudrat-Ullah, 2022). Policymakers, therefore, are introducing new regulations and designing attractive incentives to have a fair amount of cleaner energy in their overall energy supply. Consequently, with the adoption of the Paris Agreement (PA) in 2015, countries have committed, voluntarily, to the reduction of CO_2 emissions. However, it remains to be seen whether the existing policies of countries help achieve such emission reduction targets.

To reduce electricity-related CO_2 emissions, various policies and mechanisms are being pursued globally. Increasing investments and deployments of renewable and cleaner energy technologies are the most common and impactful routes to have a

[1] SDG: Sustainable Development Goals.

H. Qudrat-Ullah, *Understanding the Dynamics of Nuclear Power and the Reduction of CO_2 Emissions*, https://doi.org/10.1007/978-3-031-04341-3_1

substantial share (if not 100%) of cleaner energy in the supply mix of a country. A case in point is Germany which has set a goal of achieving 100% renewable energy by 2050 and shutting down its nuclear power plants[2] (Brown et al., 2018; Hansen et al., 2019). Solar, wind, hydro, and nuclear are the major contenders for cleaner energy supply options. There appear to be one fundamental question of concerns of the policymakers of a country: Do the existing policies will help achieve the target CO_2 reductions? If the answer is not affirmative then the next logical question worth exploring is, what would be the best (i.e., cleaner and affordable) energy supply mix to achieve the desired CO_2 emission reduction for the country? As the electricity system, today is the main source for CO_2 emissions and is widely regarded to be the key to any future decarburization pathway (Förster et al., 2012); these research questions can be pragmatically addressed in the context of the electricity system of a country.

The availability of data for a modeling-based analysis and availability of an existing dynamic model (MDESRAP) entices us to consider Pakistan's case. Pakistan, a developing country with increasing energy demand, faces challenges and opportunities in its energy sector. On the one hand, the country is facing severe issues in energy supply: (i) severe load shedding and bailouts are becoming a norm in the electricity supply sector, (ii) relatively higher prices create affordability difficultly, and (iii) electricity-related emissions are increasing at an alarming rate. On the other, the country is renewable resource-rich and has a successful nuclear power experience with an abundance of available indigenous nuclear power fuel and fuel-manufacturing capability (Altaf, 2018; WNA, 2020).

Pakistan has also signed and ratified the Paris Agreement (PA) in 2016, which requires a reduction of GHG emissions from its peak level by 2030. To achieve the reduction in GHG emissions, Pakistan has submitted its, Pakistan's Intended Nationally Determined Contribution (Pak-INDC) in 2016. The Pak-INDC presents principles and actions that are underway for reducing GHG emissions (PAK-INDC, 2016). It also describes mitigation and adaptation measures already being implemented in Pakistan together with their efficacy. However, several coal-fired power plants are being executed under China-Pakistan Economic Corridor (CPEC). Pakistan still is heavily using thermal/fossils (e.g., 67% of electricity supply was based on thermal generation [i.e., coal, oil, and gas] in 2019) for electricity generation (PEY, 2020).

Given the continuing nature of energy crises in Pakistan and the predominant nature of the fossil-based generation of electricity, it is necessary to assess the potential environmental (e.g., CO_2 emissions) impact of ongoing initiatives and policies regarding energy supply in Pakistan, a harbinger for the country's future. A modeling-based analysis, where underlying assumptions and mechanisms in the Pak-INDC are explicitly accounted for, will allow us to verify the role of current mechanisms and policies in reducing the emissions in the frame of the PA. The general question is what will be the impact of these programs on the pursuit of reducing environmental

[2] Pre-mature shutting downs of existing nuclear power plant can hamper efforts to reduce carbon dioxide and other power sector emissions and increase the cost of achieving emission reduction targets (MITEI, 2018).

emissions? As Pakistan has embraced a very ambitious nuclear power program where several new nuclear power plants (NPPs) are being built with the help of China, it is essential to relate to NPPs construction to the other measures in the energy sector of Pakistan, to see the NPPs' role in energy supply mix as well as in achieving a low carbon economy context. How do the NPPs affect the reduction of CO_2 emissions in Pakistan is a critical question to consider. This reduction of electricity-related CO_2 emissions is the central research theme of this book.

Although our focus and preference for the main dynamic modeling-based analysis will be in the context of Pakistan, at times, however, for comparative analysis, India and Saudi Arabia's cases will be evaluated and discussed as well in this book. After China, India is the second-most populous country in the world with increasing energy demand. Saudi Arabia is the world's largest producer of oil and predominantly operates with fossils-based electricity generation. Besides the geopolitical importance of these countries, all of them have ambitious nuclear power plans as a possible strategy to combat electricity-related emissions. Another commonality among these three countries is the "dominance" of thermal power in their electricity supply mix. Therefore, the term "the PIS region", in this book, will refer to these three countries: Pakistan, India, and Saudi Arabia.

The overall objective of this book is to enhance systematically our understanding of and gain insights into, the general process by which the renewables and nuclear power's role in reducing CO_2 emissions can explain the making of a low carbon economy and a systematic achievement of PA's targets about the reduction of CO_2 emissions for a country or region. By identifying and critically examining the possible energy transition pathways available for the policymakers (including of the PIS region) with the help of a dynamic modeling-based scenario approach, we attempt to contribute to the ongoing larger debate about climate change and low-carbon economic regimes. In this chapter, we begin our journey by reviewing and understanding the dynamics of power sector supply mix and CO_2 emissions, energy policies, and mechanisms including INDCs, and dynamics of nuclear power development for the countries of the PIS region.

1.2 Power Sector and CO_2 Emissions

Globally, energy-related CO_2 emissions reached a historic high of 33.1 Gt (Gigatonne) CO_2 in 2018, and the power sector accounted for almost two-thirds of the increase in total emissions (). Despite the substantial development and deployment of renewables, fossil-based generation is still the dominant source of electricity supply in the world, an alarming indicator of global policy efforts. An increasing population, unplanned growth of urbanization and industrialization, and improved lifestyle of people are the major source of the increasing electricity demand. All countries of the PIS region are facing rising demand for electricity. The electricity sector in many countries including the PIS region is still dependent on fossil-based generation. Consequently, electricity-related CO_2 emissions are substantial. Any effort to replace the fossil-based generation capacity should help the countries to reduce CO_2 emissions. Nuclear power and renewable generation present opportunities for countries that are aiming for CO_2 emissions reduction and achieving low-carbon economic regimes (Uddin et al., 2021; WNA, 2020).

Achieving target reductions in CO_2 emission and moving towards low carbon economies with renewable alone is hard to occur (Trainer, 2012, 2013). On the other hand, according to a recent annual report by IEA,[3] nuclear power remains the largest single source of low-carbon electricity generation across advanced countries (GER, 2021; WNA, 2020). Next, we present and analyze Pakistan's case along with its comparison with the other two countries of the PIS region. In particular, we provide a review and analysis of current power sector CO_2 emissions, CO_2 emissions reduction pledges (as per the INDCs of each country of the PIS region), the status of the current and planned capacity of nuclear power plants.

1.2.1 An Overview of Pakistan's Case

With an increasing population[4] with an average annual rate of over 2%, increasing although, at a slower pace, economic and industrial activities, and a severe problem of electricity theft together with troubling and consistent T&D (Transmission & Distribution) losses[5] give rise to the ever-increasing demand for electricity in Pakistan. The annual growth rate of electricity consumption is 5.6% (Qudrat-Ullah, 2022). Often, electricity supply lags the demand leading to outages and unpredictable load-shedding occurrences ranging from 2 to 8 h (for the urban area) to 2 to 18 h per day (for rural areas) (Irfan et al., 2020; Valasai et al., 2017). The economic losses due to power outages interruptions can cost between one and five percent of the GDP of developing countries (Trace, 2020). With frequent and longer electricity outages, people suffer on several fronts. With unpredictable outages, health services, especially in rural communities are negatively impacted; necessary communications are interrupted; joy-less life creates stress and anxiety.

To overcome the demand–supply gap, the Government of Pakistan has introduced several energy policies over the past three decades to attract private sector investments. Now, with the recent and ongoing implementation of CPEC-based energy projects, the electricity supply has improved, as shown in Fig. 1.1. However, these projects are mostly adding coal-fired power plants. This situation of Pakistan's electricity system can best be described as, "solving one problem (i.e., increasing the electricity supply) and falling into even a bigger problem (i.e., increased environmental emissions, increased fuel-import dependency, and increased foreign debt)." Dependency on imported fuels, which are often characterized by fluctuating and

[3] IEA: International Energy Agnecy.

[4] According to the World Bank estimates, Pakistan's population will be over 250 million by the end of 2050 (PEY, 2020).

[5] T&D losses are over 20% of total generation every year (NEPRA, 2021).

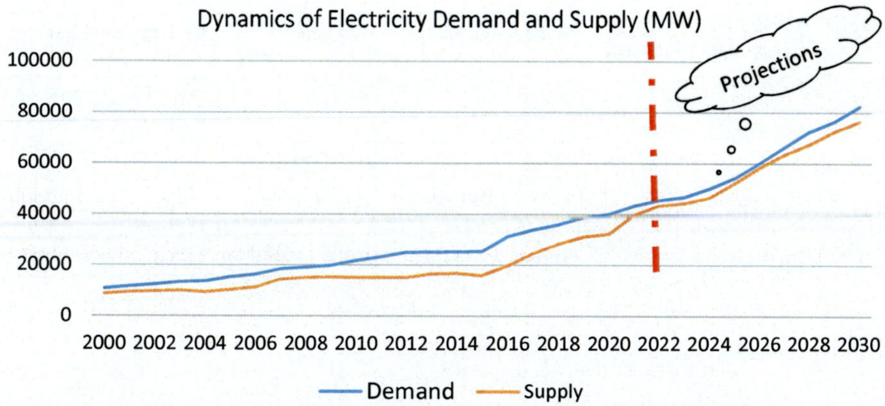

Fig. 1.1 Dynamics of demand–supply gap of electricity system of Pakistan

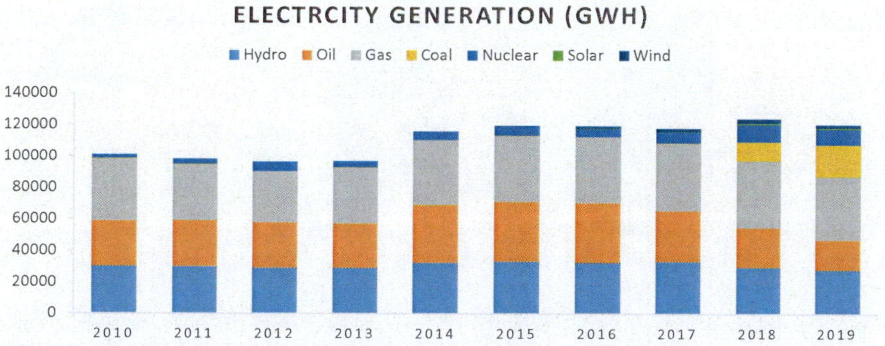

Fig. 1.2 Electricity generation in Pakistan during 2010–2019

rising prices, adds to this disturbing scenario. Therefore, at the priory, it appears plausible to be wary of the energy policy of Pakistan and conduct a critical assessment, accounting for environmental (i.e., CO_2 emissions), economic (i.e., affordability), and social (i.e., accessibility and reliability) impacts, of the Energy Policy of Pakistan.

The overall picture of electricity generation, with an installed capacity of 34,282 MW (NEPRA, 2021) for this study period, 2010–2019, is depicted in Fig. 1.2. Oil and gas-based generation remains the dominant source of generation throughout this period. Later in the year 2016 and onward, the addition of CPEC-based coal-fired power plants have been on the rise. About 74% of all CEPC-based energy projects are coal-fired power plants along with the risk of increased CO_2 emission and foreign debt (Downs, 2019). It should be noted that gas-based generation has a dominant share in the market. Most of the IPPs prefer to invest in such technologies, which can generate relatively "quick returns", often termed as the "dash-for-gas" phenomenon.

Table 1.1 Renewable energy resource potential of Pakistan

Renewable energy source	Capacity	Reference studies
Solar	29 GW	(Mirza et al., 2011; Raheem et al., 2016)
Wind	120,000 MW	(Raheem et al., 2016)
Hydro (excluding run-of-river)	42,000 MW	(Altaf, 2018; Raheem et al., 2016)
Biomass	4000 MW	(Altaf, 2018)
Geothermal	2000 MW	(Altaf, 2018)

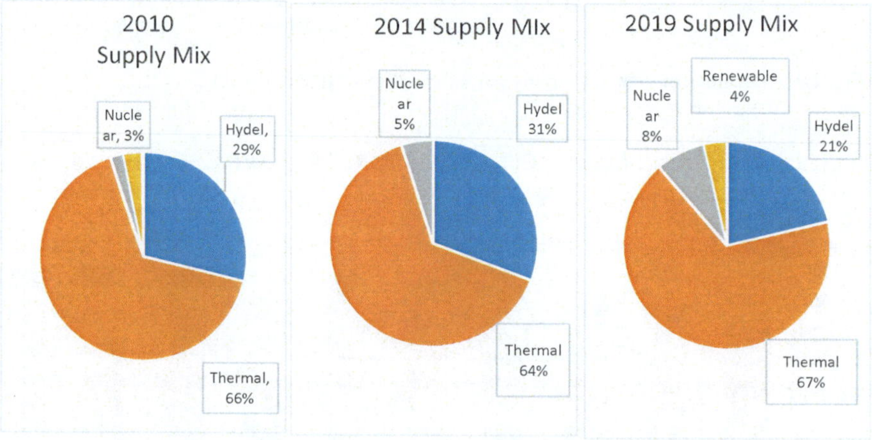

Fig. 1.3 Electricity supply mix of Pakistan

Like gas, oil-based generation also provides quick returns. Often oil-fired power plants can become operational or commissioned in one to two years. While nuclear power generation is increasing slowly but steadily, renewable generation, despite the huge potential, as listed in Table 1.1, (other than hydropower) only have started to show in the post-2015 period and now has a share of about 2% in 2019. Hydropower, although Pakistan has a huge potential of about 50,000 MW, it has hardly seen any substantial increase in its production capacity and generation during the 2010–2019 period. A simultaneous look at the electricity supply mix of Pakistan for the years 2010, 2014, and 2015, as shown in Fig. 1.3, attests to the dominant role of thermal generation in increasing electricity-related CO_2 emissions in Pakistan.

Based on the dominant share of thermal generation (oil, gas, and coal), it is reasonable to speculate relatively higher electricity-related CO_2 emissions in the electricity sector of Pakistan. Other researchers have found that renewable energy consumption has an insignificant impact on CO_2 emission in Pakistan and that, in the non-renewable energy model, natural gas and coal are the main contributors to the level of pollution in Pakistan (Lin & Raza, 2019; Rehman et al., 2020; Shah et al.,

Table 1.2 Population, GDP, and electricity consumption of the PIS region during 2010–2019

Country	Population (millions)			GDP (Billion US$)			Per capita electricity consumption (kWh)		
Years	2010	2019	GR	2010	2019	GR	2010	2019	% Change
Pakistan	179	217	2.1	177	278	5.7	506.8	671.3	33
India	1234	1366	1.1	1676	2871	7.1	667.6	1008.6	51
Saudi Arabia	27	34	2.5	528	793	5	8229.6	10,547.8	28

Data Source (OWD, 2021); GR: Average Growth Rate in percentage over the period 2010–2019; GDP is in 2010 USD (Constant)

2018). According to the World Heal Organization (WHO), the biggest global health risks are carbon emissions and pollution; approximately 3.7 million deaths occurred due to ambient air pollution in 2012, and 4.3 million deaths occurred due to residential air pollution in 2016 (Uddin et al., 2021). Thus, more emissions in Pakistan mean more health risks and a relatively higher loss of human life. Therefore, policymakers in Pakistan should revisit their energy policies and initiatives to avoid such avoidable human suffering and losses.

1.2.2 Regional Overview of Electricity Consumption and CO_2 Emissions

The overall electricity consumption in the PIS region is quite high due to the large population and increasing economic activities, as shown in Table 1.2. Although the average population growth rate of India (i.e., 1.1%) is closer to the world's population average rate of 1%, India is the fourth largest country with a population of over 1366 million in 2019. In Table 1.1, we provide an overview of the dynamics of per capita electricity consumption, capturing a first glance at the case. Compared with the year 2010, each country of the PIS region has witnessed a substantial increase in the year 2019. In the financial year, 2019, electricity consumption per capita amounted to around 109 kilowatt-hours in India. GDP of India is growing at the highest rate, over 7%, in the PIS region.

Compared with the world's average annual growth rate of about 1%, Saudi Arabia's population is growing at a higher annual rate of 2.5%. While its GDP is growing at the average rate of 5%, per capita consumption of electricity in Saudi Arabia has reached 1048 kWh, the highest in the region. The use of air-conditioning in homes, businesses, and industrial complexes, in the supper-hot weather of Saudi Arabia, appears to be a leading factor in its higher level of per capita electricity consumption. While the general assertion that "more people and more economic activity lead to more electricity consumption" still holds for the countries of the PIS region, each country's per capita electricity consumption might have various other dominant factors in play. For instance, in Pakistan's case, despite having a fairly large population and a respectable economic growth rate of 5.7%, the availability and accessibility[6] of electricity is a major cause of a relatively low consumption rate (i.e., the least among the PIS region) (Irfan et al., 2020; Valasaiet al., 2017). In addition, the highest electricity consumer in Pakistan is household with a consumption of 49% of the total electricity consumed in 2019; the other consumers are the industrial sector (26.3%), agriculture sector (9%), commercial sector (7.8%), bulk supplies (5.1%), and streetlights and others (2.8%) (PEY, 2020).

When the household sector of an economy is a major consumer of electricity, expectations, perceptions, and adaptability towards cleaner devices and appliances of the users of electricity, for example, can play a critical role in the reduction of electricity-related CO_2 emissions. Therefore, the nature of an electricity supply–demand system of a country becomes a socio-technical system. The effective management of such a socio-technical system (e.g., meeting the targets of CO_2 reductions and moving towards a low-carbon economic regime) requires decision-aiding tools to capture, in addition to dynamics of a technical system (e.g., electricity-generating power plants and associated support equipment, infrastructure, and technologies), the interactive nature of electricity users (e.g., users affordability will determine how much electricity they will consume, education, and awareness towards clean energy devices and appliances use and the adoption of renewable technologies is a major factor in the successful implementation of renewable and cleaner solutions especially in the case of remote off-grid electricity supply systems). If users and stakeholders are not engaged during the design and implementation of clean energy technologies and solutions, their successful and sustainable operations are at high risk (Qudrat-Ullah, 2022). A similar case of treating electricity demand–supply systems as socio-technical systems holds for the other two countries of the PIS region (i.e., for India and Saudi Arabia)- *another commonality* to have comparative analysis and assessment of their energy policies, CO_2 emissions, CO_2 reduction mechanism (e.g., their INDCs), and renewable and clean energy (e.g., nuclear power) in this book.

Figure 1.4 presents the dynamics of per capita CO_2 emission for each of the countries of the PIS region from the year 2010 to the year 2019. Consistent with our analysis of macro-economic factors and consumptions of electricity, Saudi Arabia, with the highest per capita electricity consumption, appears to be the biggest emitter of electricity-related CO_2 emissions in the PIS region. Compared with Pakistan with per capita CO_2 emissions of 1.15 metric tons and India with 1.91 metric tons in 2019, Saudi Arabia reached a large amount of per capita CO_2 emissions of about 17 metric tons, an alarming situation for the public and the energy policymakers of the country. It should be noted that both in Pakistan and India's case, the most populous of the PIS region, per capita CO_2 emissions, although at a slower rate, are on the rise. With rising CO_2 emissions even in 2022 (i.e., just eight years are left for these countries to meet their pledged reductions of CO_2 emission as is stated in their INDCs[7]), it is not

[6] This situation of Pakistan appears to be for away from what the UN's Sustainable Development Goal No. 7: availability of clean and affordable energy for all, requires countries to achieve.

[7] For instance, reducing GHG emissions of power plants by increasing the share of non-fossil energy to 20% is key for Iran to meet its targets in Paris Agreement (Ghadaksaz & Saboohi, 2020).

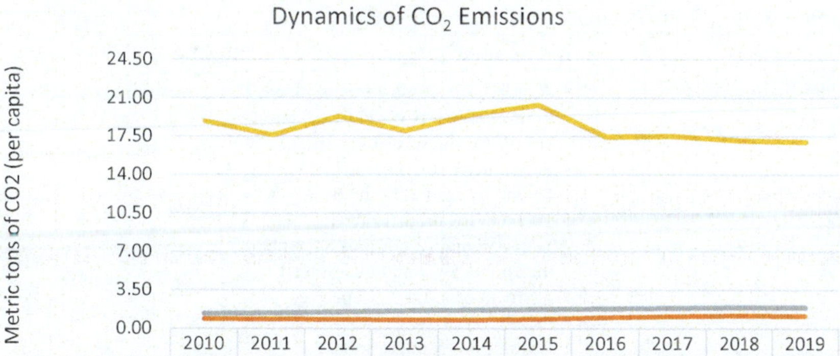

	2010	2011	2012	2013	2014	2015	2016	2017	2018	2019
Pakistan	0.86	0.84	0.82	0.79	0.79	0.85	1.00	1.12	1.17	1.15
India	1.36	1.41	1.53	1.59	1.69	1.72	1.81	1.84	1.92	1.91
Saudi Arabia	18.88	17.61	19.32	18.00	19.47	20.35	17.44	17.50	17.11	16.99

Fig. 1.4 Dynamics of CO_2 emissions in PIS region

unreasonable to suggest and commission a systematic system-wide multidimensional (e.g., affordability, reliability, and cleaner production of electricity) assessment of the existing energy policies, initiatives, and incentives of each of the countries of the PIS region.

1.3 Energy Policies and CO_2 Reducing Mechanisms of Pakistan

For the sustainable well-being, prosperity, and growth of our people on this planet, most of the nations, regions, and countries have recognized the need for the abatement of environmental emissions. Among the three major sources, electricity generation and consumption produce the most of CO_2 emissions. The other two sources are transportation and heating cooling. Like other countries, Pakistan has enacted several energy policies and initiatives for the development and deployment of renewable and cleaner electricity-generating technologies. Here we provide a brief overview and analysis of Pakistan's energy policies together with a comparative analysis of CO_2 emission reduction mechanisms of the other two countries of the PIS region.

1.3.1 Energy Policies (1990–2020) of Pakistan

Pakistan started liberalization and privatization of its energy sector in the early1990s. Since then several policies with a major focus on meeting the rising electricity demand (i.e., an average annual rate of 10% or higher). In the following Table 1.3, we provide

Table 1.3 Key Features and focus of Pakistan's energy policies

Energy policy	Key features	Key focus
1994-Energy Policy	• A bulk tariff of US cents 6.5/kWh to be charged to WAPDA (a state entity) for the sale of electricity • A premium of US cents0.25/kWh on energy soldering the first 10 years for projects above 100 MW and be commissioned by the end of 1997 and guaranteed payment of a fixed "Capacity Price"; • Exemption from corporate income tax, customs duties, sales tax, and other surcharges on imported equipment by IPPs; Simplified procedures for IPPs • Power Purchase Agreement (PPA); Fuel Supply Agreement; Foreign exchange risk insurance;	Add generation capacity
1995-HydelPolicy	• In addition to the incentivesof1994 Energy Policy, Model Implementation Agreement • A Bulk Tariff of US cents 6.1/kWh to be charged to WAPDA for the sale of electricity • The ownership of the hydropower project will be transferred to the GOP after 25 years, free of cost • Provide protection against specific force majeure risk and changes in certain taxes and duties • Guaranteed foreign exchange conversion facility and PPA for private hydropower projects	Add Hydro generation capacity
1998-Energy Policy	• The tariff will be based on open bids, In Rupees; Energy Purchase Price and Capacity Purchase Price; Guaranteed foreign exchange conversion facility; Finance raise facility • Provide protection against specific force majeure risk & changes in certain taxes and duties	Add generation capacity
2006-RE Policy	• No customs duty or sale tax for machinery meant for Renewable Energy projects; Exemption from income tax; Repatriation of equity is allowed • Parties may raise finance; Non-Muslims and non-residents shall be exempted from payment of Zakat on dividends paid by the company	Add RE generation capacity

(continued)

Table 1.3 (continued)

Energy policy	Key features	Key focus
2012-NCC Policy	• Encourage the development and promotion of hydropower generation and RE resources; Plan the necessary expansion of nuclear power; Mandate all new coal-fired power plants to perform at a high-efficiency level; • Introduce a carbon tax; Promote and provide incentives for increasing the low-carbon fossil fuels, and develop indigenous technology for CO_2 capture and storage	Add cleaner (RE & Nuclear) generation
2013-Energy Policy	• Upfront tariff and competitive bidding; Reduction of subsidies; Shift to local fuels: indigenous coal and hydro projects; The gradual-shifting of fuel-supply to IPPs; • Improve efficiency and control of electricity theft	Add local-resource based generation
2019-ARE Policy: aka the Energy Policy of Pakistan	• Have a 20% of the generation from ARE technologies by 2025 and a 30% by 2030; Lower the average basket cost of generation; Promote the exploitation of indigenous energy resources; Use open bidding for tariffs; • Create a national carbon credits trading scheme	Add RE generation capacity

Source Adapted from (Qudrat-Ullah, 2015) and (Qudrat-Ullah, 2022); ARE: Alternate Renewable Energy; NCC: National Climate Change Policy.

an overview of these policies. The majority of these energy policies (i.e., four out of seven): 1995-HydelPolicy, 2006-RE Policy, 2012-NCC Policy, and 2019-ARE Policy are focused on the promotion, development, and implementation of renewable and nuclear power generation technologies and initiatives. The key initiatives include mandating improved efficiency of all new coal-fired power plants, local production of Carbon Capture[8] (CC) technologies, and the implementation of the carbon tax.[9] However, the trajectory of electricity-related CO_2 emissions is still increasing, as is

[8] The policy initiatives regarding the local production of carbon capture technologies in Pakistan has yet to see actions on ground (Uddin et al., 2021).

[9] As of today, January 27, 2022, Pakistan has not implanted its intended carbon tax in the country. However, discussions about carbon tax are ongoing at the various levels of the government (Qudrat-Ullah, 2022).

shown in Fig. 1.1, above. This counter-intuitive situation again warrants a systematic system-wide, covering both demand and supply side of the electricity system, assessment of the Energy Policy of Pakistan (EPP).[10]

1.3.2 INDCs and CO_2 Reducing Mechanisms: A Regional Perspective

Global concerns about mitigation and adaptation of climate change and the environmental emissions led to formal actionable directives in the 1997 Kyoto Protocol-, where parties committed to a binding CO_2 reduction target. Then after several rounds of COPs (Conference of Parties) including 2007 Bali COP, 2009, and 2010 (Copenhagen & Cancun) COP, 2011 Durban COP, Warsaw (2013) and Lima (2014) COP, INDC concept was born (India-INDC, 2018). Soon after, in 1995 in Paris, countries of the world adopted an international agreement, aka the Paris Agreement, to address climate change that requires targeted emissions reduction commitments from all countries—developed and developing. In response, the PIS region countries have submitted their Intended Nationally Determined Contributions (INDC) to address climate change and achieve low-carbon economies. Table 1.4 presents each country's CO_2 reduction targets by 2030 and the mechanism to achieve those commitments.

In terms of INDC's voluntary targets, India's plan appears to be the most ambitious: a reduction of 20–25% reduction in emission intensity of GDP by 2030 compared to 2005 levels. In particular, the renewable energy target of 175 GW might face some technical and financial challenges. Only a systematic, system-wide, integrated assessment of India's Energy policies, initiatives, and incentives can provide sound insights regarding the accomplishment of these goals of CO_2 reductions.

Saudi Arabia, with its economy's predominant dependence on oil (and gas), has presented a balanced set of actions and plans regarding the reduction of CO_2 emissions, a model still to be adopted elsewhere. Besides improving energy efficiency, Saudi Arabia is trying to diversify its economy (i.e., to achieve relatively lower dependence on oil, the biggest source of CO_2 emissions in the country) and minimize the use of hydrocarbons in both upstream and downstream activities. Saudi Arabia presented their CO_2 emissions reduction strategy based on a combination of two scenarios: (i) occurrence of a diversified economy with the major support of oil-export revenues, and (ii) occurrence of accelerated domestic industrialization based on sustainable use of indecorous resources (KSA-INDC, 2015). Their INDC is focused on the implementation of the co-benefit approach.[11] The dominant share

---

[10] As the 2019-ARE Policy is the latest and existing "cleaner generation" focussed energy policy, from hereafter we will refer to it as the Energy Policy of Pakistan (EPP).

[11] Co-benefit approach: In this policy approach, the positive effects that a policy or measure aimed at one objective might have on other objectives, without yet evaluating the net effect on overall social welfare.

Table 1.4 INDC's targets and CO_2 reducing mechanisms of PIS region

Country	CO_2 targets by 2030 (INDC Statements)	Key mechanisms for CO_2 reduction
Pakistan	*"Pakistan intends to reduce up to 20% of its 2030 projected GHG emissions (*PAK-INDC, 2016*)."*	• Mitigation – implement energy efficiency measures – build expertise in developing, installing, and maintaining solar and wind power sources; promote, regulate and monitor energy efficiency; provide incentives for efficient vehicle operations; strengthen the Risk Management system; promote Energy Standards and Labeling • Adaption – upgrade and modernize rail services – upgrade and develop efficient public transport – promote Climate Smart Agriculture
India	*"India pledges a 20–25% reduction in emission intensity of GDP by 2030 compared to 2005 levels (*India-INDC, 2018*)."*	• Mitigation – Renewable Energy target 175 GW – Kochi Airport powered by solar energy – solar powered toll plazas; Swatch Bharat Mission – Delhi Metro & other MRTS; 100 smart cities – National Smart Grid Mission & Green Energy Corridor; Atal Mission for Rejuvenation &Urban Transformation; enhance energy & resource efficiency; launched Green Highways Policy; faster adoption and manufacturing of hybrid electric vehicles; vehicle fuel-efficiency standards; National Air Quality Index launched • Adaption – organic farming; efficient irrigation; watershed development; NamamiGange; National Initiative on Climate Resilient Agriculture; Climate Finance Policy

(continued)

Table 1.4 (continued)

Country	CO_2 targets by 2030 (INDC Statements)	Key mechanisms for CO_2 reduction
Saudi Arabia	"Kingdom *Saudi Arabia (KSA) seeks to achieve mitigation co-benefits ambitions of up to 130 million tons of CO_2eq avoided by 2030 annually* (KSA-INDC, 2015)."	• Mitigation – implement effective energy efficiency measures – invest in renewable energy; promote carbon capture and use; utilize natural gas – conserve, recover and reuse hydrocarbons • Adaption – encourage the reduction, recycling, and reuse of water and wastewater; develop and implement ICZM plans; develop and operationalize EWS – harness new sources of freshwater

ICZM: Integrated Coastal Zone Management; EWS: Early Warning Systems

of oil-based electricity generation is adding CO_2 emissions at the highest rate in the PIS region. Instead of using desalination plants for its water need, this country is focused on exploiting the new sources of fresh water. However, the international community still prefers explicit CO_2 reduction targets by all countries (CA, 2015). Therefore, the availability of an explicit and concrete list of projects and initiatives with associated targets of CO2 reduction would add more credibility to the plans of Saudi Arabia.

Pakistan's INDC is focused on building capacities to implement, monitor, and evaluate its various CO_2 reduction mechanisms, especially in the mitigation-relevant actions and plans. In this book, as our focus is on Pakistan's case, we will provide a detailed dynamic modeling-based assessment of Pakistan's energy policies, incentives, and initiatives. Once we have performed this analysis, we will be able to ascertain whether the existing policies of Pakistan will hit or miss the CO_2 emission reduction target as was agreed in the PA. Additionally, the best possible energy transition pathways towards a low-carbon economy will be explored.

Overall, to achieve the targeted reduction in CO_2 emissions. each country of the PIS has submitted its mitigation and adaptation initiatives and mechanisms. If these initiatives are successfully implemented and the CO_2 reduction targets as stated in their INDCs are achieved by the countries of the PIS region, it should be a celebration as a noble contribution towards sustainability and having a low-carbon planet for all of us. Cleaner energy means better health to the least, good for all of us!

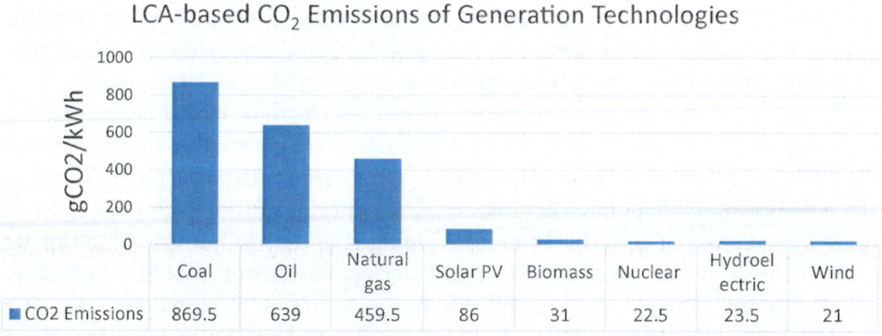

Fig. 1.5 Comparison of LCA-based CO2 emissions by various generation technologies

1.4 Nuclear Power and Electricity Supply Mix

Currently, there are about 400 nuclear power reactors that operate in about 30 countries around the globe to provide about 10% of the world's power supply (Garton et al., 2021). Nuclear power plants (NPPs) do not emit CO_2 emissions. However, one could ask how about their construction and decommissioning phases? Sure, activities involved in both phases of any power plant including a nuclear plant, do emit CO_2 emissions. Let us consider the life cycle approach, so we could see the status of CO_2 emissions across the power generation technologies. The World Nuclear Association (WNA) has conducted a comprehensive empirical study for the comparison of Life Cycle Assessment (LCA) of CO_2 emissions for various power generation technologies. This study included only those published studies that: (i) be from a credible source (e.g., published by Governments and universities), (ii) have clearly defined and used the term "lifecycle" in their assessment, (iii) include nuclear power generation and at least one other electricity generation technology, and (iv) express emissions as a function of electricity production (e.g., CO_2e/kWh or equivalent) (WNA, 2020). Figure 1.5 shows the average[12] CO_2 emissions of eight power generation technologies currently employed. These results are also consistent with IEA's estimates of environmental emissions for these technologies (IEA, 2015).

It is interesting to note that the major renewable energy source in the world, is solar power. emits more CO_2 emissions than nuclear power in about 1:4 proportion. Fossil-based generation, as was expected, is the worst environmental emissions polluter. Gas-fired power plants emit more by a factor of over 20, oil-fired power plants by a factor of over 28, and coal-based generators emit 38 times (the highest among fossils) greater than nuclear power. Biomass technology emits about 50% higher than nuclear technology. The other major two renewables, hydroelectric power, and wind power generate about the same level of emissions as an NPP. Therefore, based on these LCA-based estimates by WNA, the status of nuclear power as the cleanest source of

[12] This average is based on average of estimated by (i) nuclear specialists and (ii) other energy technologies' specialists as reported in by WNA (WNA, 2011).

electricity generation appears credible. Especially, countries with a dominant share of fossil-based generation electricity like Saudi Arabia can invest in nuclear power and lower their power sector CO_2 emissions.

Using panel datasets of 18 countries covering 95% of the global nuclear reactors, Lee et al. (2017) found that, on a long-term basis, a 1% increase in nuclear power led to a 0.26–0.32% decrease in CO_2 emissions per capita. Also, as shown in Fig. 1.6, the instrumental role of nuclear power in the avoidance of CO_2 emissions by several countries during 1971–2018. From developing countries to Europe, countries were able to avoid a total of over 60 CtCO2 during this period- another indicator of nuclear power's potential dominant role in the reduction of global greenhouse gas emissions (Garton et al., 2021). On the global front, both the International Energy Agency (IEA) and the Intergovernmental Panel on Climate Change (IPCC) have also confirmed the need for a growing role on the part of nuclear power to meet decarburization objectives such as stated and committed by countries in the Paris Agreement and consistently advocated by UN through its 17 SDGs including the one to reduce the amount of greenhouse gasses by 45% in 2030 (IPCC, 2018; IEA, 2019a; UN, 2020). IEA predicts that nuclear power use will increase to account for 15% of all annual greenhouse gas reductions by 2050 (IEA, 2019a). Therefore, for many countries especially fossil-rich economies, the role of nuclear power in the reduction of CO_2 is expected to grow.

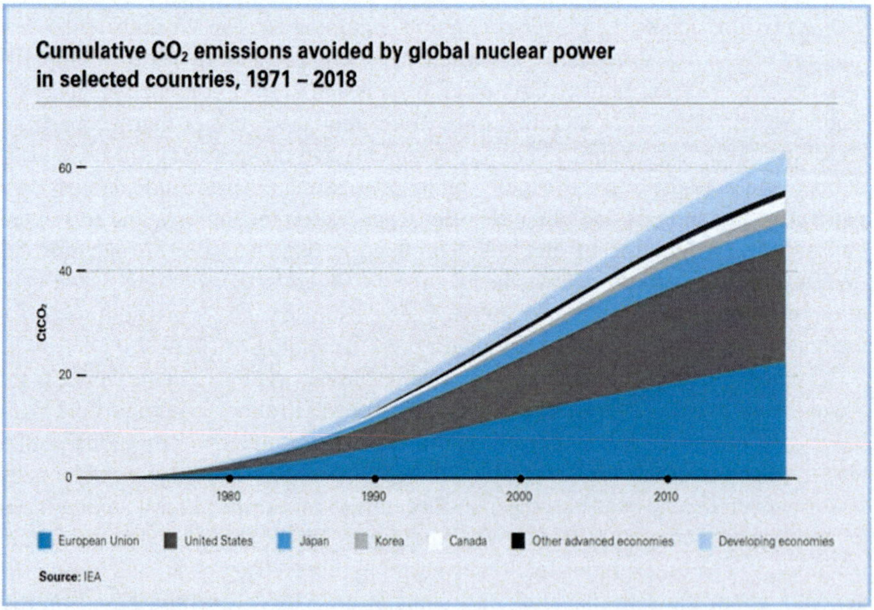

Fig. 1.6 CO_2 Emissions avoided by nuclear power in selected countries (*Source* Adopted from Garton et al., 2021)

Table 1.5 Capacity factors of various generation technologies

Energy source	Capacity factor (%)
Solar	24.9
Wind	35.4
Hydropower	41.5
Coal	40.2
Natural Gas	56.5
Geothermal	74.3
Nuclear	93.5

Source of data ONE, 2021

When we consider the well-known detrimental nature of CO_2 emissions together with a high cost of abatement of these emissions, nuclear power becomes a competitive contender of technologies for cleaner and reliable production of electricity. Thus, nuclear power is an important means of reducing greenhouse gas emissions.

Besides being one of the cleanest sources of electricity, nuclear power also commands a leading position when compared to the capacity factor of these power generation technologies. The highest capacity factor of 93.3% of an NPP is due to low maintenance and refueling (e.g., refueling is done after 1–2 years typically) shutdowns (ONE, 2021). This high capacity factor of NPPs has huge implications for fossil-based generation technologies. For instance, to generate electricity to the amount of 1GWh (a typical production rate for an NPP), we cannot have a single coal-fired or gas-fired power plant. Instead, we need to build two to three (for coal) or three to four (for gas) power plants to generate the same amount of electricity. This higher capacity for nuclear power means that an NPP can operate 93% of the time without any interruption (e.g., for maintenance and refueling), an effective means of enhancing energy production and sustainability. Thus, nuclear power is not only a clean source of power, which provides a viable means for the reduction of CO_2 emissions but also a reliable source of electricity supply. The alternate cleaner sources of power like hydro, solar, wind are weather dependent. Their intermittent nature of renewable-based electricity supply can create the issue of reliability. Also, a purely 100% renewable supply mix for the country has yet to happen (Trainer, 2012, 2013) (Table 1.5).

Another factor that is promising for the future growth of nuclear power is the invention of advanced technology-based development of the new design of the nuclear reactor, called "small modular reactor (SMR)." According to WNA and IAEA,[13] an SMR is defined as a nuclear reactor of generally 300 MWe equivalent or less capacity, designed with modular technology, pursuing economies of series of production and short construction time. The salient features of an SMR are (WNA, 2021):

- Less reliance on active safety systems, additional pumps, and AC power for accident mitigation, a key factor in enhancing the safety of SMR;

[13] IAEA: International Atomic Energy Agency.

Table 1.6 SMRs in operations

Name	Capacity	Developer
CNP-300	300 MWe[14]	SNERDI/CNNC, Pakistan & China
PHWR-220	220 MWe	NPCIL, India
EGP-6	11 MWe	at Bilibino, Siberia (cogen, soon to retire)
KLT-40S	35 MWe	OKBM, Russia

Source (WNA, 2021)

Table 1.7 SMRs under construction

Name	Capacity	Developer
CAREM25	27 MWe	CNEA & INVAP, Argentina
HTR-PM	210 MWe	INET, CNEC &Huaneng, China
ACP100/Ling long One	125 MWe	CNNC, China
BREST	300 MWe	RDIPE, Russia

Source (WNA, 2021)

- Facilitates the implementation of higher quality standards;
- Reduction of the source term as well as smaller radioactive inventory;
- Potential for sub-grade (underground or underwater) locations, a suitable technology for remote rural electrification programs;
- Multiple units on the same site;
- Suitable for remote regions and applications such as mining or desalination; and
- Easier in-situ decommissioning.

Moreover, due to their compact nature and lower power, SMRs can have relatively higher safety and protection from both natural (*e.g.* seismic or tsunami) and man-made (*e.g.* aircraft impact) risks. Their smaller and modular structure sites them for off-grid remote rural areas and can enhance the rural electrification programs of the developing countries. Several countries are actively engaged in the operations (please see in Table 1.6), development (please see in Table 1.7), and construction (please see in Table 1.8) of SMR across the globe (WAN, 2021). Relatively smaller waste materials can lead to better spent-fuel waste management. Therefore, with the increasing popularity and safety of SMRs, both developing and developed countries can harness more nuclear power for the overall reduction of environmental emissions and the provision of cleaner and cost-effective electricity to their people in urban, rural, and remote areas.

Overall, to meet one of the 17 UN's SDGs (i.e., No. 7) about the provision of affordable and cleaner energy for all with at least 80% of the world electricity must

[14] In this book, we use MW and MWe interchangeably.

Table 1.8 SMRs for near-term deployment

Name	Capacity	Developer
VBER-300	300 MWe	OKBM, Russia
NuScale	77 MWe	NuScale Power + Fluor, USA
SMR-160	160 MWe	Holtec, USA + SNC-Lavalin, Canada
SMART	100 MWe	KAERI, South Korea
VBER-300	300 MWe	OKBM, Russia
NuScale	77 MWe	NuScale Power + Fluor, USA
SMR-160	160 MWe	Holtec, USA + SNC-Lavalin, Canada
SMART	100 MWe	KAERI, South Korea
BWRX-300	300 MWe	GE Hitachi, USA
PRISM	311 MWe	GE Hitachi, USA
Natrium	345 MWe	TerraPower + GE Hitachi, USA
ARC-100	100 MWe	ARC with GE Hitachi, USA
Integral MSR	192 MWe	Terrestrial Energy, Canada
Seaborg CMSR	100 MWe	Seaborg, Denmark
Hermes prototype	< 50 MWt	Kairos, USA
RITM-200 M	50 MWe	OKBM, Russia
RITM-200 N	55 MWe	OKBM, Russia
BANDI-60S	60 MWe	Kepco, South Korea
Xe-100	80 MWe	X-energy, USA
ACPR50S	60 MWe	CGN, China
Moltex SSR-W	300 MWe	Moltex, UK

Source (WNA, 2021)

be low carbon by 2050- to achieve a realistic goal of keeping global warming within 2 °C (Qudrat-Ullah, 2022; UN, 2020). Although renewable energy is increasing its share in the overall supply of the world's energy significantly, large-scale use of solar and wind power seems to remain on the lower side, at least by 2030 (IAEA, 2020). On the other hand, nuclear power is known to be a CO_2 emission-reducing electricity generation technology. Using the panel dynamic ordinary least square method with the panel dataset consisting of 18 countries covering 95% of the global nuclear reactors, it is has been estimated that a long-term increase of 1% in the nuclear power proportion in the supply mix can lead to a 0.26–0.32% decrease in CO_2 emissions per capita (Lee et al., 2017). Therefore, to meet the need of meeting pledged reduction of CO_2 emission by 2030 (as listed in Table 1.3) and achieve the goal of having low economies with major electricity supply with low-carbon technologies, several countries planned induction and expansion of nuclear power in their electricity supply mix appears a reasonable proposition (IAEA, 2020). It is worth noting that in Europe, "going forward, nuclear and natural gas will be considered "sustainable" forms of energy" (Allen, 2022). In this book, therefore, we have set one of the objectives as to analyze and understand the dynamics of nuclear power-focused scenarios, as a major aspect of CO_2 reduction strategies, for the PIS region countries.

1.4.1 *Nuclear Power Program of Pakistan*

Pakistan has have set a very ambitious plan of building 8800 MW by 2030 and 40,000 MW by 2050 of nuclear power by 2050 (IAEA, 2020; Khurshid, 2018). Pakistan Atomic Energy Commission (PAEC) is responsible for the planning, implementation, operation, and maintenance of nuclear power plants. Pakistan is one of the four nuclear-armed states along with India, Israel, and North Korea that is not a party to the Nuclear Non-Proliferation Treaty (aka NPT) but is a member in good standing of IAEA and is a member of the World Association of Nuclear Operator (WANO) since 1989 (Ahmad, 2013).

In terms of nuclear power generation, with a mere 3% share in the supply mix in 2008, it has gained an 8% share now in 2019. Figure 1.7 portrays the dynamics of nuclear power generation in Pakistan depicting the current state of affairs. A total of 1430 MW of nuclear capacity is installed, as given in Table 1.2, comprising of five nuclear power plants. So far, Pakistan has a successful nuclear power experience with an abundance of available indigenous nuclear power fuel and fuel-manufacturing capabilities, and indigenous needed manpower. In a recent empirical study, Mahmood et al. found that nuclear power, with improved safety measures, is the way forward for Pakistan to control power-related environmental emissions (Mahmood et al., 2020). As indicated in its INDC, Pakistan is committed to reducing environmental emissions. A major task in this book project, therefore, is to assess the role of nuclear power in achieving the target of CO_2 reduction and a low carbon economy in Pakistan (Table 1.9).

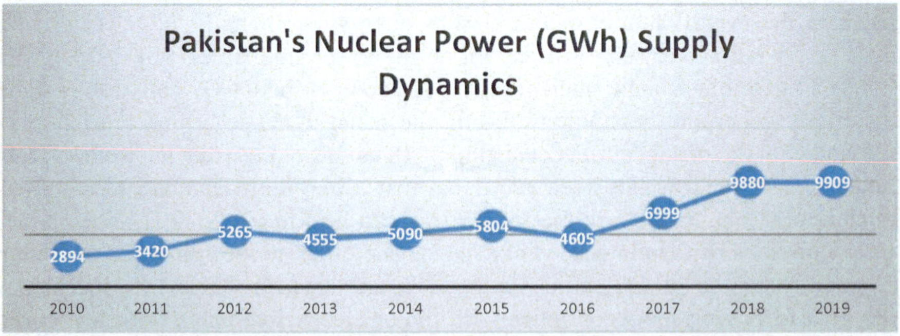

Fig. 1.7 The dynamics of Pakistan's nuclear power generation

Table 1.9 Nuclear power plant's status of Pakistan

Nuclear Power Plant	Capacity (MW)	Electricity generation in 2019 (GWh)	Human resource
1.KANUPP-1	100	175	Local
2.CHASNUPP-1	325	2294	Local
3.CHASNUPP-2	325	2448	Local
4.CHASNUPP-3	340	2694	Local
5.CHASNUPP-4	240	2298	Local
6.KANUPP-2	1040	Under construction	Local
7.KANUPP-3	1040	Under construction	Local
8.CHASNUPP-5	1000	Proposed*	Local
9.Coastal Nuclear Power Hub Baluchistan	2200	Proposed*	Local

Source (Mengal et al., 2019; Qudrat-Ullah, 2022)

Table 1.10 Dynamics of nuclear power capacity of the PIS region

Country	# of reactors in operation	Operational capacity	Under construction capacity	Planned capacity
Pakistan	6	2332 (MW)	3361 (MW)	8800 MW by 2030; 40,000 MW by 2050
India	23	6885 (MW)	5194 (MW)	30,800 (MW) by 2032–2050
Saudi Arabia	None	None	30 kW (research reactor)	17 GW by 2040

Source (IAEA, 2020; WNA, 2021)

1.4.2 A Regional Overview of Nuclear Power and Electricity Supply Mix

When it comes to nuclear power in the PIS region, although all three countries have an ambitious plan, the history and genesis of their nuclear programs are quite different. Pakistan and India are both oil-importing nations and Saudi Arabia is the largest exporter of oil in the world. Both in the case of Pakistan and India, the geopolitical and strategic objectives associated with having nuclear power capability play a major role rather than the oil prices, import-dependency, and import bills as the driving force. Saudi Arabia wants to diversify its economy and bring cleaner production in its electricity supply mix (KSA-INDC, 2015). Also, ongoing tension with Iran's development of nuclear power and associated weapon program provides a forceful push to embrace the development of nuclear technology in the country. Table 1.10 presents an overview of the dynamics of nuclear power (i.e., in terms of operational, under construction, and planned capacities) for the PIS region. In the previous (i.e., Sect. 1.4.1) we have analyzed the current and planned capacity of nuclear power plants, in the following, we will present an analysis of nuclear power both for the remaining two countries of the PIS region: India and Saudi Arabia.

India's case. In the PIS region, India has the most ambitious plan: in addition to 6885 MW of operational capacity, and 5194 MW was under-construction capacity, the proposed expansion is to build 30,800 MW capacity during the 2032–2050 period. A total of 44,720 GWh of electricity was produced by India's NPPs in 2019. The Nuclear Power Corporation of India Ltd (NPCIL) is responsible for the design, construction, commissioning, and operation of NPPs in India. When we see at India's INDC pledge to reduce the carbon emissions intensity of its economy by 33–35 percent by 2030[15] as well as increase the clean energy electricity capacity to 40 percent of the total installed capacity in the same period, the future of nuclear power in India's supply mix is going to be substantial.

Various studies have projected India's nuclear power capacity up to the year 2050. For instance, the Indian government has stated its plans, to have nuclear power equal to 25% of total generation, to be achieved by the middle of the century (WNA, 2016). A UK-base study focused on identifying the potential pathways for India for the rection of CO_2 emissions evaluated three scenarios: (i) a base case, "doing-nothing" scenario where no constraints on CO_2 emissions are in place, (ii) a first low-carbon scenario with 2.4 billions tonnes of CO_2 by 2050, and (iii) a second low-carbon scenario with the same target of CO_2 emission as in the first low carbon scenario but accounting for the India-specific local conditions (Mohan, 2016). They estimated India's total installed capacity of nuclear power in 2050 as 43 GW (in the case of reference scenarios), a total installed capacity for nuclear power of 142 GW (in the case of the first low carbon scenario), and a total installed capacity for nuclear power of 156 GW (in the case of second low-carbon scenario).

Another study (IES 47: India Energy Security Scenarios 2047, a three year short of the time horizon that other studies have considered) by the Planning Commission of India, using model-based analysis for the projections of various demand–supply pathways for India using four scenarios: (i) Least Effort Scenario, (ii) Determined Effort Scenario, (iii) Aggressive Effort Scenari, and (iv) Heroic Effort Scenario (Mohan, 2016). This studys' estimates for nuclear power capacity in India are 11.3 GW, 26.1 GW, 45 GW, and 76 GW by 2047, for each of the four scenarios, respectively.

Although these projections or estimates about India's nuclear power capacity by 2050 are divergent (e.g., by a conservative estimate it is 11.3 GW versus by an optimistic estimate of India's nuclear power capacity is about 156 GW), they do show a progressive more3 towards increasing nuclear power in the overall electricity supply mix of India. To realize this estimated fleet of nuclear power plants, various factors like local versus foreign supplied reactors, the establishment of a reliable supply

[15] Some researchers have pointed out the limitations of some of the India's mechanism to reduce CO_2 emissions. For instance, "Green India Mission" is expected to deliver 50–60% of its planned output (Sundriyal & Dhyani, 2015). Thus, the need for cleaner energy production including of nuclear power in India becomes even more pronounced.

chain of related requirements and technologies, political will and commitments, safety concerns, and cost and financing challenges have to be effectively considered and addressed by the government of India. Sooner these issues are resolved, the addition of this ambitious fleet of nuclear power plants will put India on track not only to ave the targeted reduction of CO_2 emissions but also begin the journey towards having a low-carbon economic regime.

Given the large population (i.e., 4th largest country in the world) and rising electricity demand, as is shown in aforementioned Table 1.10, this ambitious nuclear power plan is not that surprising. Moreover, during the past two decades, India's electricity supply mix, as is portrayed in Fig. 1.8, was predominantly based on coal-fired power plants with a share of 67% in 2010 and 72% in 2019. Besides ongoing efforts and push for renewables, India's ambitious nuclear power plan appears to have a greater role in the reduction of electricity-related CO_2 emissions in India. Only a substantial reduction in coal-fired power generation together with replacement by cleaner production (e.g., with renewables and nuclear power) of energy can help India to achieve its goal of having a low-carbon economy.

Saudi Arabia's case. In the case of Saudi Arabia, oil and gas-fired power plants are the major sources of electricity generation. In the overall electricity supply mix of Saudi Arabia, about 60% total electricity supply is generated by burning oil (ITA, 2018). Saudi Arabia is ranked 10th in the list of the largest greenhouse gas emitting country in the world (IEA, 2019b). Therefore, it is not unreasonable to think that a reduction in electricity-related CO_2 emissions is needed and is a strategic decision.

In a similar assertion in the case of Saudi Arabia, according to Prince Turki al-Faisal, the shift into conservation, renewables, and nuclear energy is linked to growing energy demand (Drollette Jr, 2016). Meeting the growing energy demand and reducing the rising co2 emissions, appear to justify such a multi-pronged strategy of Saudi Arabia In terms of nuclear power, the government of Saudi Arabia has already signed some agreements with nuclear-technology suppliers like Argentina and Korea. For instance, for the development of small-scale nuclear power plants, Saudi Arabia has signed for CAREM-25, a 27MWe NPP, with Argentina, and a SMART-100, a 100 MWe NPP, with Korea (WNA, 2019). For large capacity NPPs such as APR-1400 and OPR-1000 with KEPCO, Korea (IAEA, 2018). Also, Saudia Arabia's alternate energy development agency, KA-CARE, has signed an agreement with KAERI, KORA for a possible building of small NPPs such as SMART-100 (WNA, 2019).

In a recent study using model-based scenario analysis, Esmail and Cheong (2021) have evaluated various possible pathways for CO_2 reduction in Saudi Arabia. They evaluated three scenarios (Esmail & Cheong, 2021):

(i) Base scenario: It is assumed all operational power plants in the country will continue as usual until the end of their lives without either any rehabilitation or introduction of any new power plants. Essentially, it is a "do-nothing" scenario.

(ii) Rehabilitation scenario: It is *assumed that old would be rehabilitated and with needed retrofitting efficiency will be enhanced.* Scenario.

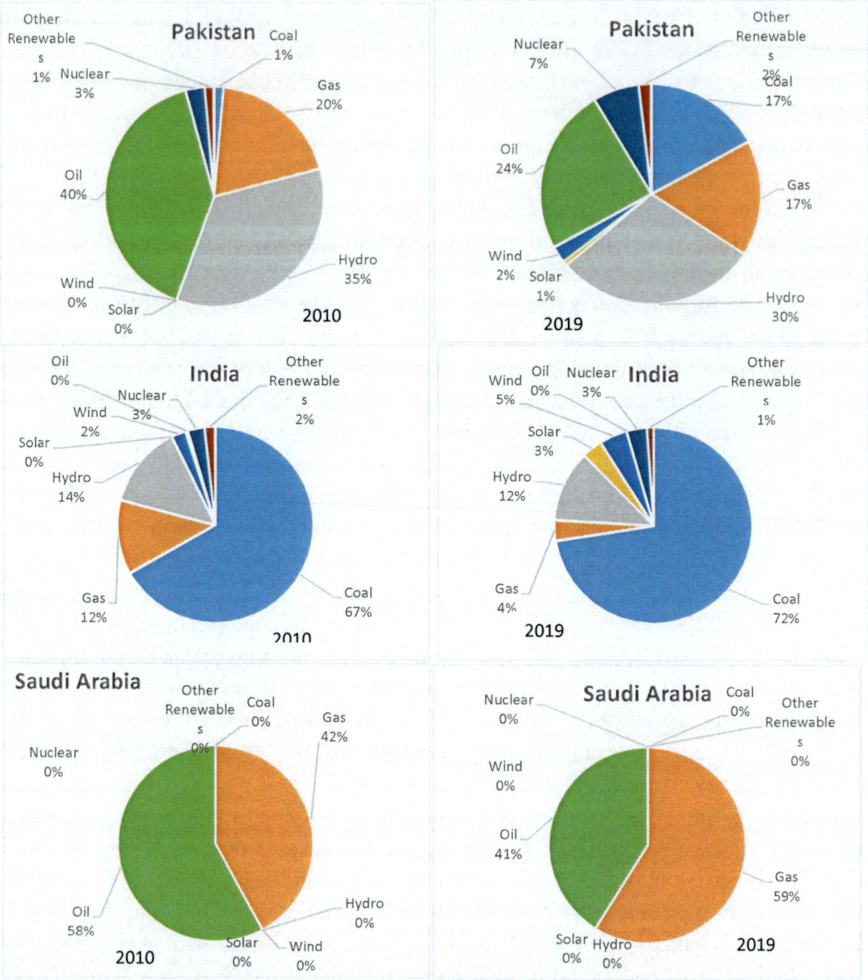

Fig. 1.8 Dynamics of electricity supply mix of the PIS region for 2010 and 2019 years

(iii) Clean electricity production scenario: In this scenario, it assumed that older than 30 years of age power plants will be replaced with renewables (e.g., Saudi Arabia has the best quality wind and sun resources in the world(Faruqui et al., 2011; IRENA, 2012, 2018; WNA, 2019; Zarrouk & Moon, 2014) and nuclear power (e.g., Saudi Arabia already selected three potential sites for nuclear power pants: (i) Jubail on the Arabian Gulf in East region, (ii) Tabuk in West region) close to the Red Sea, and (iii) and Jizan on the Red Sea in South region (WNA, 2019). technologies in the electricity supply mix of Saudi Arabia.

Based on the assessment of these scenarios, Esmail and Cheong (2021) used the model, MESSAGE,[16] provided various insights and conclusions regarding supply of electricity and the potential reduction in electricity-related CO_2 emissions and as a way forward for Saudi Arabia, as follows:

(i) Business as a usual case (i.e., the reference scenario) can lead to potential blackouts (i.e., the shortages of electricity) after the year 2025.
(ii) Rehabilitation of existing power plants (i.e., in the Rehabilitation scenario) can delay the possible occurrence of blackouts by ten years.
(iii) The introduction of new alternative energy resources including nuclear power is needed for a sustainable supply of electricity in Saudi Arabia.
(iv) The solar, advanced gas, wind, advanced steam, and advanced combined cycle would be the top five major alternate energy technologies in the future (i.e., the Clean electricity production scenario).
(v) It is expected that installed capacity-wise ranking, nuclear power stand at the sixth position by 2050.
(vi) The combined (i.e., APR-1400[17] and SMART[18]-100) nuclear fleet would show the highest installed capacity, followed by APR-1400 only and SMART-100 only.
(vii) The calculated LUEC[19] for the combined nuclear power option exceeded the LUEC for the APR-1400 only scenario by approximately 35%.
(viii) The introduction of the carbon tax will make solar power the best (cheaper) electricity generation technology, followed by nuclear power (i.e., the 2nd best) in Saudi Arabia.
(ix) Due to the implementation of CO_2 tax, nuclear power showed an 8 to 74% increase in the installed capacity due to CO2 taxation during the period from 2025 to 2050 (Esmail & Cheong, 2021).

Based on this study, there are two major takeaways regarding potential nuclear power development in Saudi Arabia: (i) considering the existing LUEC, a combined nuclear fleet appears to work best, (ii) implementation of a carbon tax will increase the feasibility of nuclear power in Saudi Arabia (Esmail & Cheong, 2021).

Overall, with regards to nuclear power, Saudi Arabia's ambitious plan appears plausible. With no operational or under-construction nuclear power plant, almost all of its electricity generation with fossils (oil and gas-fired power plants) as is shown in Fig. 1.8, and the highest per capita CO_2 emissions in the PIS region, setting up an ambitious plan of having 17 GW of nuclear power by 2040 appears a step forward. Diversifying its electricity supply mix with cleaner generation in any proportion will

[16] MESSAGE: Model for Energy Supply Strategy Alternatives and their General Environmental Impact, developed by the International Institute for Applied Systems Analysis (IIASA) in Austria since the 1980s (ENE, 2009; Messner & Strubegger, 1995).

[17] APR-1400: Advanced Power Reactor 1400 MW electricity, is an advanced pressurized water nuclear reactor designed by Korea Electric Power.

[18] SMART-100: System-integrated Modular Advanced Reactor, is an integrated-type small reactor, developed by KAERI (Korea Atomic Energy Research Institute, Korea.

[19] LUEC; Levelized Unit Energy Cost.

certainly help in the rising electricity-related emissions in Saudi Arabia. It should be noted that without the strong infrastructure and manpower, which is needed for the successful realization and operations of nuclear power plants, Saudi Arabia's nuclear power plan might take longer than planned. However, the successful acquisition, installation, and operation of a nuclear power plant (The Barakah NPP[20]) by the neighboring country, UAE, can trigger faster development of nuclear power projects in Saudi Arabia.

As the focus of this book is on Pakistan's case, Pakistan, with item-wise safeguard agreements with IAEA (WNA, 2021), is making notable progress towards the realization of its planned additional nuclear capacity. Although China is providing technical support, Pakistan still faces some logistical and financial issues about the building of planned nuclear power plants. In terms of manpower and needed infrastructure, Pakistan is self-sufficient (WNA, 2019). If Pakistan could add 8800 MW by 2030, as planned, then one could hope to see the long-term goal of building a larger share of nuclear power in its electricity supply mix as realizable. Therefore, using dynamic modeling-based analysis, we will assess Pakistan's case in this book project in two respects:

(i) do the current energy policy, various initiatives and incentives help Pakistan to achieve the pledged CO_2 emission by the year 2030, and
(ii) what is the best electricity supply mix for Pakistan to move towards a low carbon economic regime?

The complexity of these questions should be recognized. For instance, in the case of the first question, instead of thinking about a dichotomous answer (i.e., yes or no), a systematic multidimensional (i.e., economic (e.g., dynamics of electricity prices), environmental, (e.g., electricity-related CO_2 emissions) and social (accessibility and adoption of clean energy solutions by the users) assessment is required. Likewise, effective evaluation of an existing energy policy or design of an alternate one should account for the interactions among the key aspects of the electricity demand–supply system (e.g., the dynamics of demand, generation capacity, investments, regulations, supply mix, profitability, electricity prices, and policy incentives). Therefore, using a dynamic model-based[21] scenario approach, we will attempt o address these questions in this book project. Besides providing the answers and policy insights to the decision makers we hope to contribute to the larger and ongoing debate about the role of nuclear power in the mitigation of climate change and global warming challenges across the world.

[20] The Barakah NPP is the United Arab Emirates' first nuclear power station, the first nuclear power station in the Arabian Peninsula, and the first commercial nuclear power station in the Arab World. It consists of four APR-1400 nuclear reactors.

[21] The development and validation of dynamic model to be used in this book is described in the Chapters 4 and 5 of this book.

1.5 Summary of Chapter 1: The Key Insights

In this chapter, we have set the objective of this book project, "understanding the dynamics of nuclear power and electricity-related CO_2 missions." While the focus is on Pakistan's case, as the dynamic modeling-based analysis to be presented in this book will draw on Pakistan's case data, a systematic comparative overview of the PIS region (i.e., Pakistan, India, and Saudi Arabia) in terms of its power sector emissions, electricity supply mix, CO_2 reducing policy initiative and mechanisms, and nuclear power plans is also carried out. Figure 1.9 presents our theoretical framework for policy assessment and design that is being used in this book. This design consideration in a policy design requires the policymaker to apply a systematic and holistic decision-making approach where the achievement of reliable, affordable, and cleaner electricity supply is the goal of their energy policy. This framework is generic enough to apply to the energy policy assessment and design exercise for any country of the PIS region and beyond.

Key policy insights are:

- Despite being rich in renewable sources (e.g., hydro, solar, and wind), Pakistan's electricity supply still, in 2019–2020, is predominantly thermal (i.e., coal, oil, and gas) based primarily because of IPPs preferred investments in "quick-return-generating" oil, coal, and gas-fired power plants.
- When compared with India, and Saudi Arabia, per capita electricity consumption, despite having a relatively higher population growth rate, is the lowest in Pakistan because of availability, accessibility, and reliability issues of its electricity supply system.

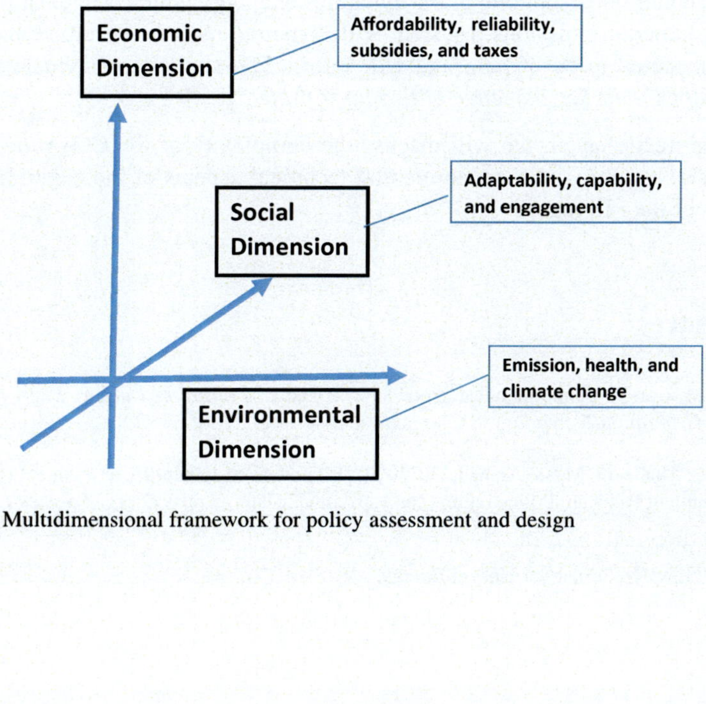

Fig. 1.9 Multidimensional framework for policy assessment and design

- Saudi Arabia's per capita electricity is the highest among the countries of the PIS region mostly due to air-conditioning used to manage work and living in super-hot weather.
- India's per capita electricity consumption is rising due to increased economic activities and the improved lifestyles of its people requiring urgent attention from their energy policymakers and planers towards its thermal-dominated electricity supply mix.
- Saudi Arabia's electricity-related CO_2 emissions are the highest in the PIS region.
- With the existing LCUE,[22] a combined nuclear power plants' fleet appears feasible for Saudi Arabia.
- The introduction of a CO_2 emissions tax will make nuclear power prices competitive in all the countries of the PIS region.
- The electricity-related CO_2 emissions of Pakistan are increasing due to the dominance of thermal generation in its electricity supply mix.
- The electricity-related CO_2 emissions of India are increasing due to the dominance of coal-fired generation in its electricity supply mix.
- The electricity-related CO_2 emissions of Saudi Arabia are increasing due to the dominance of oil and gas-fired power generation in its electricity supply mix.
- Given the dominant share of thermal generation in the electricity supply mix, rising CO_2 emissions of the PIS region, and to ascertain whether the CO_2 reductions targets as committed by these countries in their INDCs, a systematic and system-wide assessment of their energy policies appears to be imminent.
- Although the nuclear power plans of Pakistan, India, and Saudi Arabia are progressive and ambitious, bringing the actual operational capacity of NPPs into the electricity supply mix appears challenging primarily due to financial difficulties and constraints.
- For a better understanding of the dynamics of energy policies, nuclear power, and environmental emissions, a system-wide dynamic-modeling exercise and analysis are expected to be of help and will inform and support policymakers in their decisions—*the key objective of this book project.*

In the next chapter, we will discuss the complexity of the CO_2 reduction task in terms of various socio-economic and technical aspects of the electricity supply systems of the PIS region.

References

Ahmad, M. (2013). Scope of nuclear power in Pakistan. *Journal of Nuclear Energy Science & Power Generation Technology*, S1. https://doi.org/10.4172/2325-9809.S1-001. Accessed 17 July 2021.

Akbar, M., Thaheem, M., & Arshad, H. (2017). Life cycle sustainability assessment of electricity generation in Pakistan: Policy regime for a sustainable energy mix. *Energy Policy, 111*, 111–126.

[22] LCUE: Levalized Cost per Unit of Energy.

Ali, R., Kuriqi, A., & Kisi, O. (2020). Human-environment natural disasters interconnection in China: A review. *Climate, 8*(48), 1–28. https://doi.org/10.3390/cli8040048

Allen, N. (2022). *Europe changes tack—Natural gas and nuclear now "green"*. https://seekingalpha.com/news/3795105-europe-changes-tack-natural-gas-and-nuclear-now-green?mailingid=26574846&messageid=2900&serial=26574846.10818&utm_campaign=rta-stock-news&utm_content=link-3&utm_medium=email&utm_source=seeking_alpha&utm_term=26574846.10818. Accessed 2 Feb 2022.

Altaf, M. (2018). Renewable energy; identifying potential and development of solar energy to contribute to sustainable energy delivery in Pakistan. *International Journal of Engineering and Advanced Technology, 7*(3), 28–33.

Brown, T. W., Bischof-Niemz, T., Blok, K., Breyer, C., Lund, H., & Mathiesen, B. V. (2018). Response to 'Burden of proof: A comprehensive review of the feasibility of 100% renewable-electricity systems.' *Renewable and Sustainable Energy Reviews, 92*, 834–847.

Climate Analytics (CA). (2015, November 24). *The paradox of Saudi Arabia's climate plans*. https://climateanalytics.org/latest/the-paradox-of-saudi-arabias-climate-plans/. Accessed 27 Jan 2022.

Drollette Jr., D. (2016, January 2). View from the inside: Prince Turki al-Faisal on Saudi Arabia, nuclear energy and weapons, and Middle East politics. *Bulletin of the Atomic Scientists*, 19.

Downs, E. (2019). *China-Pakistan economic corridor power projects: Insights into environmental and debt sustainability*. https://www.energypolicy.columbia.edu/research/report/china-Pakistan-economic-corridor-power-projects-insights-environmental-and-debt-sustainability. Accessed 23 July 23 2021.

ENE. (2009, April 26). *Energy modeling framework: Model for energy supply strategy alternatives and their general environmental impact (MESSAGE)*. International Institute for Applied Systems Analysis. http://www.iiasa.ac.at/Research/ENE/model/message.html. Accessed 27 Mar 2021.

Esmail, M., & Cheong, H. (2021). Studies on optimal strategy to adopt nuclear power plants into Saudi Arabian energy system using MESSAGE Tool. *Science and Technology of Nuclear Installations, 2021*, 1–26. https://doi.org/10.1155/2021/8818479

Faruqui, A., Hledik, R., & Wikler, G. (2011). *Bringing demand-side management to the Kingdom of Saudi Arabia final report*. Brattle Group.

Förster, H., Healy, S., Loreck, C., Matthes, F., Fischedick, M., Lechtenböhmer, S., Samadi, S., & Venjakob, J. (2012). *Information for policy makers 2: Analysis of the EU's energy roadmap 2050 scenarios* (SEFEP Working Paper 2012).

Garton et al. (2021). *Why small modular reactors will shape the future of the nuclear debate*. https://www.whitecase.com/publications/insight/why-small-modular-reactors-will-shape-future-nuclear-debate. Accessed 29 Aug 2021.

Global Energy Review (GER). (2021). *Global energy review 2021*. IEA's Annual Report. https://www.iea.org/reports/global-energy-review-2021. Accessed 27 Jan 2022.

Ghadaksaz, H., & Saboohi, Y. (2020). Energy supply transformation pathways in Iran to reduce GHG emissions are in line with the Paris Agreement. *Energy Strategy Reviews, 32*, 100541.

Hansen, K., Mathiesen, B., & Skove, I. (2019). Full energy system transition towards 100% renewable energy in Germany in 2050. *Renewable and Sustainable Energy Reviews, 102*, 1–13.

International Atomic Energy Agency (IAEA). (2018). *e database on nuclear power reactors*. International Atomic Energy Agency. https://pris.iaea.org/PRIs/

IAEA. (2020). *Country nuclear power profiles 2020 edition*. Pakistan. https://www-pub.iaea.orgMTCD/publications/PDF/cnpp2020/countryprofiles/Pakistan/Pakistan.htm. Accessed 12 Jan 2021.

IEA. (2015). *Energy technology perspectives*. OECD/IEA.

IEA. (2019a). *Global energy and CO2 status report 2018*. https://iea.blob.core.windows.net/assets/23f9eb39-7493-4722-aced-61433cbffe10/Global_Energy_and_CO2_Status_ Report_2018.pdf. Accessed 12 Mar 2021.

International Emissions Agency (IEA). (2019b). *CO2 emissions from fuel combustion highlights* (2019b ed.). International Emissions Agency.

India-INDC. (2018). *India's intended nationally determined contributions India-(INDC)—Towards climate justice*. Government of India Ministry of Environment, Forest and Climate Change. http://moef.gov.in/wp-content/uploads/2018/04/revised-PPT-Press-Conference-INDC-v5.pdf. Accessed 26 Aug 2021.

IPCC. (2018, October). *Special report: Global warming of 1.5°C*. IPCC. www.ipcc.ch/sr15

International Renewable Energy Agency (IRENA). (2012). Renewable energy technologies: Cost analysis series. *Power Sector, 1*(4–5).

International Renewable Energy Agency (IRENA. (2018). *Renewable power generation costs in 2018*. International Renewable Energy Agency.

Irfan, M., et al. (2020). Assessing the energy dynamics of Pakistan: Prospects of biomass energy. *Energy Reports, 6*, 80–93.

International Trade Administration (ITA). (2018). Saudi Arabia country commercial guide, Saudi Arabia power. *export.gov*. https://www.export.gov/apex/article2?id=SaudiArabia-Power. Accessed 22 Aug 2021.

Khurshid, J. (2018). Small Modular Reactors (SMRs)–A future nuclear power option. *Csinsight, 1*(2), 65–82.

KSA-INDC. (2015). *The intended nationally determined contribution of the Kingdom of Saudi Arabia under the UNFCCC*. https://www.ctc-n.org/sites/www.ctc-n.org/files/UNFCCC_docs/ksa-indcs_english.pdf. Accessed 29 Aug 2021.

Lee, S., Kim, M., & Lee, J. (2017). Analyzing the impact of nuclear power on CO2 emissions. *Sustainability, 9*, 1428. https://doi.org/10.3390/su9081428

Lin, B., & Raza, Y. (2019). Analysis of energy-related CO2 emissions in Pakistan. *Journal of Cleaner Production, 219*, 981–993.

Mahmood, N., Danish, Wang, Z., & Zhang, B. (2020). The role of nuclear energy in the correction of environmental pollution: Evidence from Pakistan. *Nuclear Engineering and Technology, 52*(6): 1327–1333.

Manisalidis, I., Stavropoulou, E., Stavropoulos, A., & Bezirtzoglou, E. (2020). Environmental and health impacts of air pollution: A review. *Frontiers in Public Health, 8*, 14. https://doi.org/10.3389/fpubh.2020.00014

Mengal, A., Mirjat, H., Walasai, D., Khatri, A., Harijan, K., & Kaili, A. (2019). Modeling of future electricity generation and emissions assessment for Pakistan. *Processes,7*(212), 1–26.

Messner, S., & Strubegger, M. (1995). *User's guide for MESSAGE III*. International Institute for Applied Systems Analysis.

Mirza, A., Ahmed, S., & Khalil, S. (2011). Renewable energy on Pakistan: Opportunities and challenges. *Science Vision, 6*(17), 13–20.

MIT Energy Initiative (MITEI). (2018). *The future of nuclear energy in a carbon-constrained world*. An Interdisciplinary MIT Study. https://energy.mit.edu/wp-content/uploads/2018/09/The-Future-of-Nuclear-Energy-in-a-Carbon-Constrained-World.pdf. Accessed 22 Jan 2022.

Mohan, A. (2016, August). *The future of nuclear energy in India*. Technical Report (ORF Occasional Paper #98). https://www.orfonline.org/research/the-future-of-nuclear-energy-in-india/. Accessed 27 Jan 2022.

National Electric Power Regulatory Authority (NEPRA). (2021). *State of industry report 2020*. https://nepra.org.pk/publications/State%20of%20Industry%20Reports/State%20of%20Industry%20Report%202020.pdf. Accessed 24 Aug 2021.

NRDC. (2021). *Fossil fuels: The dirty facts*. https://www.nrdc.org/stories/fossil-fuels-dirty-facts. Accessed 20 Aug 2021.

Office of Nuclear Energy (ONE). (2021). *Nuclear power is the most reliable energy source and it's not even close*. https://www.energy.gov/ne/articles/nuclear-power-most-reliable-energy-source-and-its-not-even-close. Accessed 24 Aug 2021.

OWD. (2021). *Our world in data*. https://ourworldindata.org/energy/country/saudi-arabia. Accessed 22 Aug 2021.

PAK-INDC. (2016). *Pakistan's intended nationally determined contribution*. Ministry of Climate Change, Government of Pakistan. http://mocc.gov.pk/PublicationDetail/ZGM3YjM3MDEtMjcxOS00MDNjLTliMTMtNTk2MmNhZDRiMWMw. Accessed 17 Aug 2021.

PEY. (2020). *Pakistan energy yearbook 2019*. Ministry of Petroleum and Natural Resources.

Qudrat-Ullah, H. (2015). Independent power (or pollution) producers? Electricity reforms and IPPs in Pakistan. *Energy, 83*(1), 240–251.

Qudrat-Ullah, H. (2022). A review and analysis of renewable energy policies and CO_2 emissions of Pakistan. *Energy, 238*(Part B), 121849.

Qudrat-Ullah, H., Akrofi, M., & Kayal, A. (2020). Analyzing actors' engagement in sustainable energy planning at the local level in Ghana: An empirical study. *Energies, 13*, 1–29.

Raheem, A., Abbasi, S. A., & Memon, A. (2016). Renewable energy deployment to combat energy crisis in Pakistan. *Energy, Sustainability* and *Society, 6*(1). https://doi.org/10.1186/s13705-016-0082-z

Rehman, U., Cai, Y., Siyal, A., Mirjat, H., Fazal, R., & Kashif, R. (2020). Cleaner and sustainable energy production in Pakistan: Lessons learnt from the Pak-TIMES model. *Energies, 13*(108), 1–21.

Shah, Z., Hou, F., & Mirza, F. M. (2018). The role of renewable and non-renewable energy consumption in CO_2 emissions: A disaggregate analysis of Pakistan. *Environmental Science and Pollution Research International, 25*(31), 31616–31629. https://doi.org/10.1007/s11356-018-3059-y

Sundriyal, C., & Dhyani, P. (2015). Significance of India's INDC and climate justice: An appraisal. *Current Science, 109*(12), 2187–2187.

Trace, S. (2020). *South Africa's crippling electricity problem*. https://www.opml.co.uk/blog/south-Africa-s-crippling-electricity-problem. Accessed 29 July 2021.

Trainer, T. (2012). Can Australia run on renewable energy? The negative case. *Energy Policy, 50*, 306–314.

Trainer, T. (2013). Can Europe run on renewable energy? A negative case. *Energy Policy, 63*, 845–850.

Uddin, R., Shaikh, A. J., Khan, H. R., Shirazi, M. A., Rashid, A., & Qazi, S. A. (2021). Renewable energy perspectives of Pakistan and Turkey: Current analysis and policy recommendations. *Sustainability, 13*, 3349. https://doi.org/10.3390/su13063349

United Nations (UN). (2020). *Policy briefs in support of the high-level political forum 2020*. Accelerating SDG 7 Achievement in the time of COVID-19. https://sustainabledevelopment.un.org/content/documents/26235UNFINALFINAL.pdf. Accessed 13 Jan 2021.

Valasai, D., Uqaili, A., Memon, R., Samooe, R., Mirjat, N., & Harijan, K. (2017). Overcoming electricity crisis in Pakistan: A review of sustainable electricity options. *Renewable and Sustainable Energy Reviews, 72*, 734–745.

World Nuclear Association (WNA). (2011). *Comparison of Lifecycle greenhouse gas emissions of various electricity generation sources*. WNA Report. http://www.world-nuclear.org/uploadedFiles/org/WNA/Publications/Working_Group_Reports/comparison_of_lifecycle.pdf. Accessed 27 Aug 2021.

World Nuclear Association (WNA). (2014). *Nuclear power in Pakistan*. http://www.world-nuclear.org/info/Country-Profiles/Countries-O-S/Pakistan/. Accessed 20 July 2021.

World Nuclear Association (WNA). (2016). *Nuclear Power in India*. http://www.world-nuclear.org/info/Country-Profiles/Countries-G-N/India/. Accessed 13 June 2020.

World Nuclear Association (WNA). (2019). *Nuclear power in Saudi Arabia*. World Nuclear Association. https://www.world-nuclear.org/information-library/countryprofiles/countries-o-s/saudi-arabia.aspx

World Nuclear Association (WNA). (2020). *Nuclear power in the world today*. https://www.world-nuclear.org/information-library/current-and-future-generation/nuclear-power-in-the-world-today.aspx, Accessed 15 Jan 2021

World Nuclear Association (WNA). (2021). *Small nuclear reactors*. https://www.world-nuclear.org/information-library/nuclear-fuel-cycle/nuclear-power-reactors/small-nuclear-power-reactors.aspx. Accessed 29 Aug 2021.

Zarrouk, S., & Moon, H. (2014). The efficiency of geothermal power plants: A worldwide review. *Geothermics, 51*, 142–153.

Chapter 2
Understanding the Nature of CO_2 Emissions Reduction Task

2.1 Introduction

Setting up the ambitious goals of CO_2 reduction in general and from the electricity supply sector of a country in particular, is "easier said done." The task of CO_2 reduction is a complex, dynamic task. The complexity of the task of CO_2 reduction primarily stems from the socio-economic, environmental, and technical nature of electricity supply systems. For instance, although electricity generation, transmission, distribution, and consumption are tasks of technical nature, their impact on the users is often social (e.g., communication, healthcare, and well-being of all the strata of users), economic (e.g., affordability of electricity for the industrial, commercial, and residential users), and environmental (e.g., emissions, air pollution, and disruptive situations for the natural habitat).

Existence of a multitude of uncertainties (e.g., fuel prices, fuel availability, effects of both man-made and natural hazards, electricity demand, regulation and standards, innovative and disruptive technologies, and effects of environmental emissions) together with several non-linearities (e.g., the relationships between electricity prices and its industrial and commercial demand is rarely proportional) and delays (e.g., physical delays in the construction of power plants and delays in the arrival of machinery; information delays such as in the approval of various licensing and permits related to building, operation, and dismantling activities of power plants) adds to the complexity CO_2 reduction task.

The energy system has complex interactions with other sectors of the economy of a region or country. For instance, the provision of modern healthcare services to the largely rural population of developing countries is heavily dependent on the reliable and affordable availability of electricity; fossil-based generation can cause carcinogenic diseases leading to the increased healthcare bill for people and the governments;

H. Qudrat-Ullah, *Understanding the Dynamics of Nuclear Power and the Reduction of CO_2 Emissions*, https://doi.org/10.1007/978-3-031-04341-3_2

SMEs,[1] the backbone of any economy, can only thrive if they are provided with uninterrupted access of energy sources; renewable energy solutions, especially for the remote off-grid areas, require the development of local technical capacity; issues around the adaptability to the modern application and energy-efficient devices; and dual-nature of the use of energy (Qudrat-Ullah et al., 2020) in some cities (e.g., use of charcoal for cooking and electricity for lighting and other needs) are some situations of the dynamic interactions among a country's economic, social, and environmental sectors.

Public attitudes (e.g., towards nuclear power or wind power "not-in-my-backyard" attitude) and the varying nature (often conflicting) of the objectives of the stakeholders of the electricity system (e.g., while IPPs[2] prefer to invest in "quick-return" generation technologies (e.g., oil and gas-fired power plants), governments and consumers want more renewable energy in the supply mix) further complicate the task of CO_2 reduction for the policymakers. Therefore, all the efforts and initiatives aimed at the system-wide reduction of CO_2 emissions (e.g., to meet a target) and building a low-carbon economy for a region or a country should systematically and holistically account for these factors, which provide a glimpse of dynamic interactions among energy, economy, and environmental sectors.

2.2 The Complexity of CO_2 Reduction Task

The CO_2 reduction to meet the targets (e.g., as agreed by the countries in the PA) and achieve a low-carbon economy is a noble but complex task. Involvement of various stakeholders (e.g., the government at various levels, regulators, IPPs, and consumers) and often with conflicting objectives (e.g., while IPPs' prefer to invest in a quick-return electricity generation technology like oil and gas-fired power plants, consumers and environmentalist want more of renewable generation), socio-economic, and technical nature of electricity systems and the existence of several uncertainties make the task of CO_2 reduction a difficult and complex task. The dynamic interactions between various variables of the electricity system (e.g., demand, generation capacity, investments, and regulations) add to the complexity of the task—the reduction of electricity-related CO_2 emissions.

2.2.1 Socio-Economic and Technical Nature of Electricity Systems

Electricity generation, transmission, and distribution are highly technical tasks. For instance, at the outset, building a dam for hydropower or installation of solar/wind

[1] SMEs: Small and Medium Enterprises.

[2] IPPs: Independent Power Producers.

power plants, the proven and mature sources of renewable energy, are highly technical tasks in nature. However, in most dam-building cases, a huge and unpleasant displacement of humans and natural habitats has to occur. Especially, the impact of a large dame (e.g., the Three Gorges Dam in China [Fu et al., 2010]) has both known (e.g., resettlement of people) and unknown (e.g., in case of an accident, the impact on the life and welfare of downstream people and habitat would be difficult to estimate) impact and social implications. In addition, the success of off grid solar energy initiatives is often dependent upon the acceptance and adaptability of the community people (Qudrat-Ullah et al., 2020)—a complex social issue.

Dealing with people's sensitivities to electricity-related environmental emissions is predominantly a social issue. Likewise, the price of electricity is directly linked to the consumers' affordability aspect,—another social issue. Often the conflicting objectives of various stakeholders (e.g., local community, business community, regulatory agencies, and governments) further complicate the electricity generation, transmission, distribution, and consumption decisions. Table 2.1 shows the involvement of various actors and stakeholders in a variety of decisions including regulations and incentives, investments, operations of power plants, pricing of electricity, and purchasing and procurements. With the prevalence of such a multitude of social and technical aspects, the electricity system of a country is essentially a socio-technical system. The task of reduction of CO_2 emissions has become a complex task. Without accounting for these social, economic, environmental, and technical factors, any energy policy or initiative to reduce electricity-related CO_2 emissions can hardly succeed (Qudrat-Ullah, 2022).

2.2.2 *Existence of Uncertainties and Non-Linearities*

All socio-economic and technical systems such as electricity systems[3] face several uncertainties that make energy policy decisions difficult. Figure 2.1 shows eight critical uncertainties including (i) Energy Demand, (ii) Fuel Prices, (iii) Fuel Availability, (iv) Effects of Environmental Emissions, (v) Regulations and Standards, (vi) Natural Hazards, (vii) Man-made Hazards, and (viii) Technological Disruptions, that are commonly associated with most of the energy systems across the globe. Understanding the nature and the potential impacts of these uncertainties is essential for the design, development, and assessment of any energy policy.

i. *Energy Demand*. The dynamics of energy demand are multi-dimensional in nature. For instance, energy demand for heating and cooling buildings is driven by a climatic component (e.g., temperature and humidity of the area as is in the case of Saudi Arabia-higher temperature require more energy for cooling), a socio-economic component (e.g., population density and behavior of people, gross domestic product, price of energy as is the case of Pakistan and India—higher population drives energy demand), and by a technological component

[3] In this book, the terms electricity systems(s) and energy system(s) are used interchangeably.

Table 2.1 Stakeholders and nature of decisions in an electricity system

	Generation		Transmission		Distribution		Consumption	
Stakeholders/Nature of decisions	Government	IPPs	Government	Private firms	Government	Private firms	Government	Consumers
• Regulatory	✓		✓		✓		✓	
• Investment	✓	✓	✓	✓	✓	✓	✓	
• Operational	✓	✓	✓	✓	✓	✓	✓	✓
• Pricing	✓				✓	✓	✓	
• Purchasing	✓			✓		✓		✓
• Environmental	✓	✓	✓	✓	✓	✓	✓	✓

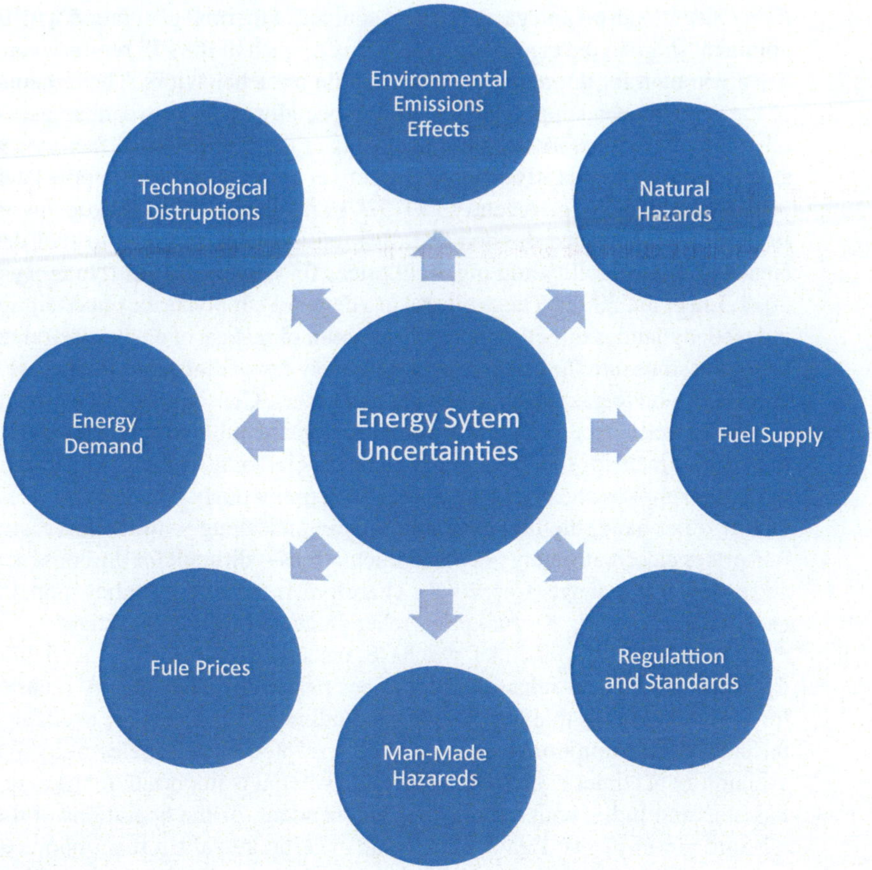

Fig. 2.1 Energy system uncertainties

(e.g., design and material determining the thermal properties of the building, efficiency of heating and cooling systems—older buildings and inefficient appliances and devices consume more energy) (Isaac & van Vuuren, 2009; Trotter et al., 2009). In a recent empirical study about the dynamics of energy demand due to climate change, Deroubaix et al. (2021), show that "*overall continental areas, the climate-driven energy demand trends for heating and cooling were weak, changing by less than 10% from 1950 to 1990, but become stronger from 1990 to 2030, changing by more than 10%*" (Deroubaix et al., 2021, p. 1). Any energy policy assessment and design exercise that aims to achieve a substantial reduction in electricity-related CO_2 emissions, therefore, should account for these demand uncertainties and variabilities.

ii. *Fuel Prices*. Both oil and gas, the major sources of thermal generation with the dominant share in the electricity supply mix of each of the PIS region's counties are known for unpredictable and volatile price behaviors. The dynamics of the gap between supply and demand, especially in swear weather cases or other supply (e.g., wars or major accidents at the fuel resource facility e.g., attack on Saudi Arabia's Rastanora Oil Refinery disrupted the oil supply [IOD, 2011]) or demand (e.g., recently, COVID-19 pandemic-based lockdowns and restrictions suppressed the demand) disruptive events can lead to fuel deficiency or unavailability and high fuel prices for power systems (Hu & Ryan, 2017; Liu et al., 2009). These oil and gas demand-supply, price uncertainties-induced, dynamics directly impact the amount of release of electricity-related CO_2 emissions into the environment- relatively fewer emissions in the case of a shift from oil to gas use by the power generators. Consequently, the growing share of electricity production from natural gas combined with the scarcity of firm contracts for fuel procurement creates electricity-supply vulnerability when gas supplies and prices are fluctuating continuously (Hu & Ryan, 2017). On the other hand, the fuel availability dynamics along with the forecasting fuel prices can create energy demand fluctuations—difficult for the consumers and provider to manage (Cec, 2008). Therefore, in any energy policy modeling endeavor, accounting for fuel price-related uncertainties is essential.

iii. *Fuel Availability/Supply*. As we have discussed above, fuel availability dynamics are closely related to fuel prices behaviors. The "energy security" for a country entails that fuel supplies including power plants are available at the point of consumption to ensure an uninterrupted supply of electricity. Fuel availability becomes even more vulnerable when it is imported; in the case of Pakistan and India, both are oil-import dependent for the operations of their oil-fired power plants. Heavy reliance on uncertainty-ridden fuel imports can provide additional and powerful incentives for countries, as is the case for Pakistan and India, to develop renewable and nuclear power capacity. Policy-makers can hardly overlook the role of fuel availably in electricity generation and associated CO_2 emissions (Qudrat-Ullah, 2022), therefore, the need for the development and promotion of indigenous renewable and clean energy sources become even more pronounced.

iv. *Effects of Environmental Emissions*. Although the uncertainties of environmental emission's damages are far too large, various studies have attempted to quantify them (Spadaro & Rabi, 2008). For instance, thermal power generation, the dominant source of electricity in the PIS region is a major cause of CO_2 and other greenhouse gas (GHG) emissions into the atmosphere. These emissions are responsible for trapping the heat of the sun in the lower layers of the atmosphere—which would otherwise escape into space—thereby generating global warming (Barreira et al., 2017). Although several studies have tried to quantify the health impacts of these emissions (e.g., in 2016, a study analyzed the impacts of emission from 257 coal-fired power plants in the European Union [EU] and found (i) more than 22,900 premature deaths, (ii) 21,000 hospital admissions, (iii) 11,800 cases of chronic bronchitis in adults and

51,700 cases of bronchitis in children, and (iv) 23,502,800 restricted activity days and 6,575,800 workdays lost [Hansen, 2016]), the estimation of the actual magnitude of the socio-economic and environmental impacts of these emissions is still a work in progress. Huge uncertainties exist about the effects of emissions on the weather system of our planet. The economic and societal effects of global warming, extreme weathers, longer than normal drought spells, more frequent occurrence of floods are hard to quantify. However, the global community, almost unanimously, is focused on the mitigation and adaptation strategies to combat these rising CO_2 emissions. Energy policy-makers when they want to design and implement policies are expected to account for such uncertainties related to electricity-related CO_2 emissions.

v. *Regulations and Standards.* Over the past three decades, various countries have embraced "privatization" aka "liberalization" of their energy sector. Despite the variety of approaches that countries have adopted (e.g., Feed-in-Tariff (FIT), RPS,[4] and PPAs[5]) usually the common dominator is the same. That is, these regulations are driven by (i) meeting the rising electricity demand (e.g., most of the developing countries are faced with the challenge to close the demand-supply of energy) and (ii) international regulations (e.g., the PA,[6] INDCs,[7] and UN's SDG[8]s) on energy use and its effects on climate change and global warming. In the case of developing countries (e.g., Pakistan and India), the intended goals of enacting regulations and standards to attract private investments include setting up attractive incentives for IPPs. (e.g., the bulk purchase agreements in the case of Pakistan). Effects of these regulations are mixed, at best. For instance, most of the IPPs in Pakistan has invested in "quick-return" generating oil and gas-based generations and are the leading electricity-related CO_2 emissions producers in the country (Qudrat-Ullah, 2015). On the other hand, developed nations like Canada and Germany have successfully enacted FIT-based policies to develop and substantially increase the share of renewable energy in their supply mix (Necoechea-Porras et al., 2021).

However, as they say, "no one model fits all," energy sector regulations have also brought in some notable unintended consequences, e.g., litigations. For instance, despite the initial period of optimism and success, by 1998 IPPs representing two-thirds of private power capacity contracts were plagued with allegations of corruption, technical inconsistencies, and attempts by the government to renegotiate tariffs (Fraser, 2005). In contrast, IPPs have had a less successful early history in India—with Enron's attempt to set up an IPP widely cited as the failure of India's 1991 effort to open up the power

[4] The Renewable Portfolio Standards (RPSs)-based energy policies provide an explicit target for renewable energy (e.g., 30% renewable energy by 2025 or 10% Solar by 2015).

[5] PPAs: Power Purchase Agreements.

[6] PA: The Paris agreement.

[7] INDCs: Intended Nationally Determined Commitments.

[8] SDGs: Sustainable Development Goals.

generation sector (Mukherjee, 2014). However, later with more reforms and regulations in India, have enabled IPPs to add much-needed generation capacity in India. India should capitalize on the policies that have brought this success so far and seek innovative ways to improve these important gains.

Overall, frequent changes in a country's regulations and standards related to the energy sector are a major cause of concern for investors and IPPs. Specifically, if the intended goal of an energy policy is to bring in more renewable generation to combat the electricity-related CO_2 emissions, as is the case with current energy policies of all three countries of the PIS region, stability and continuity of regulation and incentives, not the uncertainties, in a balanced manner is critical and essential.

vi. *Natural Hazards.* Natural disasters including earthquakes, hurricanes, tornados, avalanches, volcanic eruptions, ice storms, landslides, floods, etc. badly affect the energy systems. They not only physically damage the power generation and transmission infrastructure but also can cause severe and long power blackouts and outages (Chang et al., 2007). Only with a solid backup plan and preparation (e.g., scenario-based planning for emergence preparedness and alternate ready-to-use sources of energy), the effects of natural disasters-based uncertainties can be minimized.

vii. *Man-made Hazards.* Terrorism, war, nuclear power plant accidents, and airplane crashes are examples of man-made disasters (Oral & Dönmez, 2010). Be it is a natural or man-made disaster, the priority of the decision makers is to provide or re-establish the interrupted supply of electricity. Therefore, ongoing monitoring, evaluation, and updating of power plant-specific safety systems and measures are key to avoiding a site-specific man-made hazard. To minimize the post-disaster damage and electricity supply disruption, again a sold alternate plan, ready-to-implement should be prepared. Diesel-based power generation is commonly used as a power generation source in a disaster-struck area. Depending upon the magnitude and duration of the use of such a backup power generation system, additional CO_2 is expected to be produced.

viii. *Technological Disruptions.* The power sector globally is facing several technological challenges. Innovative technologies are becoming a norm in today's fast-paced world. Smart grids to new energy storage devices, these technological disruptions can severely impact investments. Costly retrofitting (e.g., adding technology for the conversion of coal-production to clean-coal production) or installation of new technologies, say for monitoring and control of electricity production-related environmental emissions is becoming common and is highly unpredictable. On the other hand, "significant shifts are taking place with the increased deployment of renewable technologies and the rapid development of on-site generation, information, and control technology. PV panels, batteries, and demand appliances, when combined, can help reduce the reliance on the grid and the utilities' generation capacity" (Fuentes et al., 2019, p. 4). However, the integration of these intermittent and asynchronous energy supplies with the grid system and utilities can become challenging for developing countries. These technological innovations have multidimensional implications for the energy policy decisions maker, For example, how equitable is the implementation of FITs where several consumers of the neighborhood are facing affordability challenges? How would the sustainable development of the area be affected by decreasing reliance on the grid? Would the government's subsidies to promote distributed renewable energy be sustainable? These socio-economic issues associated with these new and emerging distributed energy solutions need to be addressed and accounted for in the design of a renewable and clean energy-focused energy policy. Overall, these technological disruptions are expected to be in favor of a more renewable and clean energy supply resulting in reduced CO_2 emissions.

The relationships among several variables of an electricity system are non-linear, thus adding to the complexity of electricity systems. Consider the relationship between an operator's overtime work and her productivity—a highly nonlinear; in the beginning, her productivity can increase (e.g., due to learning) but if s/he does not take a rest and continues to overwork for long then her productivity will fall or even complete collapse, the *burn-out phenomenon*, can occur. The productivity of the experienced power plant operators rarely follows a proportional path: more experience leads to increased productivity but after some time productivity reaches a plateau. Likewise, when the price of electricity decreases, its industrial usage can see some growth. However, after a while, when even the price continues to fall, industrial usage of electricity will saturate (e.g., because the production reaches its maximum capacity)—non-linear relationship.

In another example, the relationship between the sustained relative oil price and the indicated share of production capacity of oil is non-linear, as is exhibited in Fig. 2.2. This relationship is identified based on the historical data of Pakistan's case. The extreme points of the graph represent the limit of substitution. For some consumers, depending upon their electricity-dependent application, it may be impossible to switch over despite the significantly more attractive price of the substitute. For example, the price of electricity may become much cheaper (e.g., more than six times) than the price of oil, but certain consuming industries (e.g., transport sector)

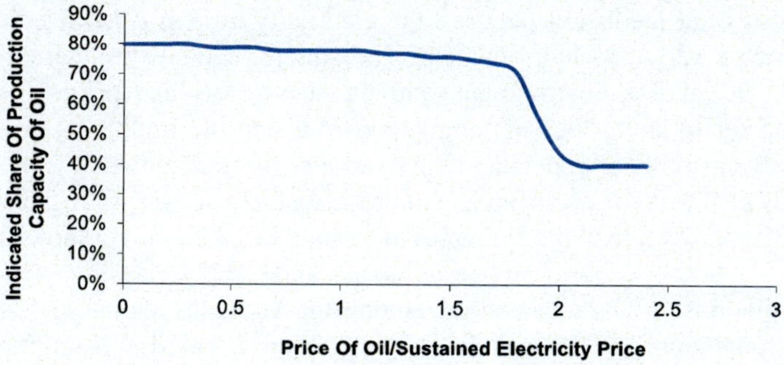

Fig. 2.2 Share of indicated production capacity of oil

will still be constrained to substitute for electricity. Similarly, certain consuming appliances (e.g., hospital equipment) will be dependent upon electricity regardless of how inexpensive the substitute (oil) is.

Such nonlinear relationships abound in sociotechnical systems such as energy systems. Therefore, the utility of policy-supporting analysis of an energy system but without an explicit representation and modeling of its critical nonlinear relationships is not of much use for the policymakers (Qudrat-Ullah, 2016).

Overall, the existence of these uncertainties, non-linearities, and delays and their potential effect on the electricity generation system of a country adds to the complexity of the task of the policymaker, which is to ensure a cleaner and sustainable supply of electricity. Therefore, for an effective policy assessment and design initiative aimed at the reduction of CO_2 emissions, an explicit recognition, and accounting of these uncertainties, delays, and non-linearities together with the understating of the energy systems as socio-economic and technical systems needs to be developed.

2.3 Key Dynamics of Electricity Supply System of the PIS Region and CO_2 Emissions

Figure 2.1 presents the dynamics of population, Gross Domestic Product (GDP), industrial and residential electricity consumption of three countries of the PIS region. The dynamics of GHG emissions associated with the economies of the PIS region are shown in Fig. 2.3. Looking at these two figures simultaneously, it would be naïve to ignore the complex interactions and feedbacks among these variables (e.g., population, GDP, and consumption of electricity) of an electricity system. For instance, the rising gap between electricity demand and supply of any country creates an ongoing need for additional generation capacity. Depending upon the specifics of the regulatory regimes and policy incentives (e.g., the preferential electricity purchasing rates, tax incentives, simplified procedures, etc.), IPPs and other investors can add new capacity. However, meeting the need (i.e., reliable availability of electricity) of electricity with indigenous resource-based generation technologies (e.g., hydropower) often conflicts with the quick-return-seeking preferences of the investors. Thus, the interactions (and feedback) between the electricity system's variables, over time, in and across various sectors demand, investment, capital (production capacity), resource, production, environment, and finances (costs and prices) needs to be accounted for in any policy-making exercise. Using the traditional correlation or trend analysis, to better understand the dynamics of electricity and CO_2 emissions, can hardly deal with the underlying feedback structures of the system that are responsible for the production of the dynamics of various variables, as are shown in Figs. 2.3 and 2.4.

The relationship and interactions among the variables shown in Figs. 2.3 and 2.4 (i.e., population, GPD, and the varying sectorial-based consumption of electricity) cannot be assumed as linear. A linearized transformation of the non-linear

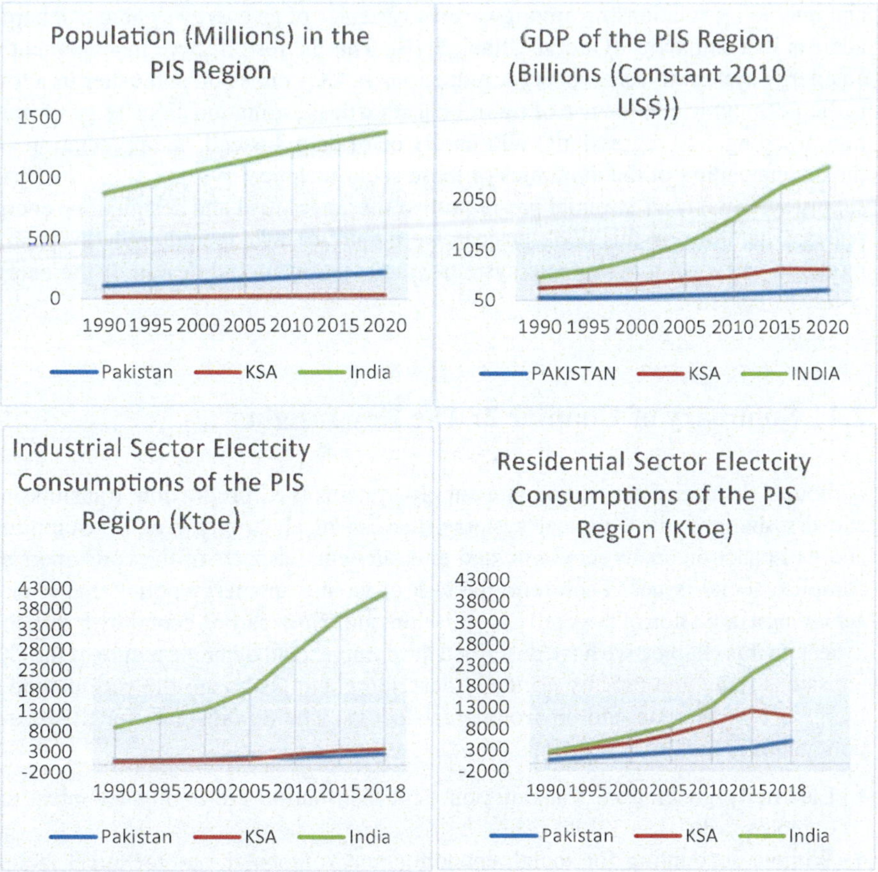

Fig. 2.3 Dynamics of population, GDP, and electricity consumption in the PIS region (*Data Sources* World Bank Indicators and IEA)

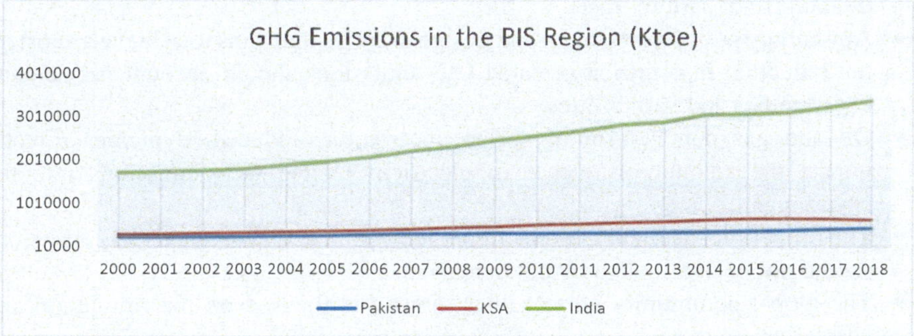

Fig. 2.4 Dynamics of GHG emissions in the PIS region

and interactive relationship among various variables of an energy system can hardly address its complexity (Qudrat-Ullah, 2016). That is, for effective management of an energy system to achieve target reductions in CO_2 emission and usher in a low-carbon economic era, the use of traditional statistical estimation alone (e.g., correlation, trending, and regressions) will hardly be of help. Instead, an appreciation and the understanding of the dynamics of these socio-technical systems (i.e., electricity supply systems) is an essential pre-requite of the assessment and design of an energy policy. Therefore, in this book project, we intend to develop and apply a dynamic model to carry out a systematic system-wide assessment and design of the energy policy of Pakistan.

2.4 Summary of Chapter 2: The Key Insights

Although electricity generation and supply systems (i.e., production, transmission, and distribution) are technical systems, the use of electricity (i.e., consumption) and its impact on socio-economic and environmental aspects of the consumers are complex, social issues. Therefore, the task of having an energy policy that aims at achieving a reduction of targeted CO emissions and a low-carbon economy is difficult, at best. In this chapter, we have described the complex and dynamic nature of the CO_2 emissions reduction task by critically describing and analyzing the various underlying socio-economic and environmental factors. The key insights for the energy policymakers are:

- Electricity generation, transmission, and distribution are highly technical and complex tasks.
- Without accounting for social, economic, environmental, and technical factors, any energy policy or initiative to reduce electricity-related CO_2 emissions is bound to fail.
- Electricity systems face several uncertainties that make energy policy decisions difficult if not hard to tract.
- Any energy policy assessment and design exercise that aims to achieve a substantial reduction in electricity-related CO_2 emissions should account for demand uncertainties and variabilities.
- Oil and gas demand–supply, price uncertainties-induced, dynamics directly impact the amount of release of electricity-related CO_2 emissions into the environment,
- The role of fuel availably in electricity generation and associated CO_2 emissions should not be overlooked by policymakers.
- The global community almost unanimously is focused on the mitigation and adaptation strategies to combat these rising emissions.
- Frequent changes in a country's regulations and standards related to the energy sector are a major cause of concern for private investors and IPPs.

- With a solid backup plan and preparation (e.g., scenario-based planning), the effects of natural disasters-based uncertainties can be minimized.
- Costly retrofitting (e.g., adding technology for the conversion of coal-production to clean-coal production) or installation of new technologies, say for monitoring and control of electricity production-related environmental emissions is becoming highly unpredictable.
- The utility of policy-supporting analysis of an energy system but without an explicit representation and modeling of its critical nonlinear relationships is of limited use for policymakers.
- Using the traditional correlation or trend analysis, to better understand the dynamics of electricity and CO_2 emissions, can hardly account for the underlying feedback structures of the system that are responsible for the production of the dynamics of various variables.
- An appreciation and the understanding of the dynamics of these socio-technical systems (i.e., electricity supply systems) is an essential pre-requite of the assessment and design of an energy policy aimed at achieving cleaner electricity production and a low-carbon economy.

Overall, the existence of uncertainties, non-linearities, the dynamic interactions among the variables of an electricity system, and their potential effect on the electricity generation system of a country adds to the complexity of the task of the policymaker, which is to ensure a cleaner and sustainable supply of electricity. Therefore, a dynamic modeling-based analysis that accounts for these complexity-adding factors and sheds light on the causal nature of interactions among the variables of an energy system can provide useful policy insights to the decision-makers in the energy policy domain.

References

Barreira, A., Patierno, M., & Bautista, C. (2017). Impacts of pollution on our health and the planet: The case of coal power plants. *Perspective, 28*, 1–10. https://wedocs.unep.org/bitstream/handle/20.500.11822/22218/Perspective_No_28_web.pdf. Accessed 2 Sep 2021.

Commission for Environmental Cooperation (CEC). (2008). *Renewable energy as a hedge against fuel price fluctuation*. http://www.cec.org/files/documents/publications/2360-renewable-energy-hedge-against-fuel-price-fluctuation-en.pdf. Accessed 23 Jan 2022.

Chang, S. E., McDaniels, T. L., Mikawoz, J., & Peterson, K. (2007). Infrastructure failure interdependencies in extreme events: Power outage consequences in the 1998 ice storm. *Springer Natural Hazards, 41*, 337–358.

Deroubaix, A., Labuhn, I., Camredon, M., Gaubert, B., Monerie, P.A., Popp, M., Ramarohetra, J., Ruprich-Robert, Y., Silvers, L.G., & Siour, G. (2021). Large uncertainties in trends of energy demand for heating and cooling under climate change. *Nature Communications, 12*, 5197. https://doi.org/10.1038/s41467-021-25504-8

Fraser, J.M. (2005). *Lessons from the independent private power experience in Pakistan* (Energy and Mining Sector Board Discussion Paper No. 15). World Bank Group.

Fu, B.J., Wu, B.F., Lü, Y.H., Xu, Z.H., Cao, J.H., Niu, D., Yang, G.S., & Zhou, Y.M. (2010). Three Gorges Project: Efforts and challenges for the environment. *Progress in Physical Geography:*

Earth and Environment, 34(6), 741–754. https://doi.org/10.1177/0309133310370286. Accessed 2 Sep 2021.

Fuentes, R., Hunt, C., Lopez-Ruiz, H., & Manzano, B. (2019). The "iPhone effect": The impact of dual technological disruptions on electrification. *Competition and Regulation in Network Industries, 21*(2), 110–123.

Hansen, J. (2016). *Europe's dark cloud—How coal-burning countries are making their neighbors sick*. WWF, Climate Action Network (CAN) Europe, Health and Environment Alliance (HEAL), and Sandbag. https://env-health.org/IMG/pdf/dark_cloud-full_report_final.pdf. Accessed 3 Sep 2021.

Hu, D., & Ryan, S. (2017). Quantifying the effect of natural gas price uncertainty on economic dispatch cost uncertainty. In *The Proceedings of the 2017 IEEE Power & Energy Society General Meeting (PESGM)*. https://doi.org/10.1109/PESGM.2017.8274496. Accessed 2 Sep 2021.

International Oil Daily (IOD). (2011). *Houthi Drones Hit Aramco's Riyadh Refinery*. https://www.energyintel.com/pages/eig_article.aspx?DocId=1100977. Accessed 2 Sep 2021.

Isaac, M., & van Vuuren, D. P. (2009). Modeling global residential sector energy demand for heating and air conditioning in the context of climate change. *Energy Policy, 37*, 507–521.

Liu, C., Shahidehpour, M., Fu, Y., & Li, Z. (2009). Security-constrained unit commitment with natural gas transmission constraints. *IEEE Transactions on Power Systems, 24*(3), 1523–1536.

Mukherjee, M. (2014). *Private participation in the Indian Power Sector: Lessons from two decades of experience*. Directions in Development: Energy and Mining, World Bank Group.

Necoechea-Porras, P., López, A., & Salazar-Elena, J. (2021). Deregulation in the energy sector and its economic effects on the power sector: A literature review. *Sustainability, 13*, 3429. https://doi.org/10.3390/su13063429

Oral, B., & Dönmez, F. (2010). The impacts of natural disasters on power systems: Anatomy of the Marmara earthquake blackout. *Acta Polytechnica Hungarica, 7*(2), 107–118.

Qudrat-Ullah, H. (2015). Independent power (or pollution) producers? Electricity reforms and IPPs in Pakistan. *Energy, 83*(1), 240–251.

Qudrat-Ullah, H. (2016). *The physics of stocks and flows of energy systems*. Springer.

Qudrat-Ullah, H. (2022). A review and analysis of renewable energy policies and CO_2 emissions of Pakistan. *Energy, 238*(Part B), 121849.

Qudrat-Ullah, H., Akrofi, M., & Kayal A. (2020). Analyzing actors' engagement in sustainable energy planning at the local level in Ghana: An empirical study. *Energies, 13*(2020), 1–29.

Spadaro, V., & Rabi, A. (2008). Estimating the uncertainty of damage costs of pollution: A simple transparent method and typical results. *Environmental Impact Assessment Review, 2*, 166–183.

Trotter, I. M., Bolkesjø, T. F., Féres, J. G., & Hollanda, L. (2009). Climate change and electricity demand in Brazil: A stochastic approach. *Energy, 102*, 596–604.

Chapter 3
Energy Policy Instruments for the Promotion of Renewable and Clean Energy

3.1 Introduction

Global concerns regarding climate change and CO_2 emissions are well documented (Manisalidis et al., 2020; Uddin et al., 2021). As a result, energy policymakers across the world are putting efforts and are taking initiatives to mitigate and reduce CO_2 emissions. often they have to re-assess and re-design their energy policies. A core consideration and primal issue in the design of any energy policy is what kinds of policy incentives should be incorporated in the policy document that attracts and protects investments by foreign as well as local investors. Renewable energy and nuclear power technologies are capital intensive. Both Pakistan and India's energy projects need large investments. On the other hand, IPPs and private investors, in addition to the quick return on their investments, require tax incentives, simplified procedures (e.g., in the approvals and permits), power purchase and fuel supply guarantees, currency hedging, etc. Depending on the objectives of a county, each policy will have specific items and rules to address the relevant issues. Each country of the PIS region has its specific needs in terms of energy accessibility, affordability, and cleanliness. No single set of policy instruments can fit all countries of the PIS region. Some countries want investments in hydropower that typically take a longer time to recoup the invested capital as compared with fossil-based generation (e.g., quick dash for gas: gas-based generation is quickly installed and monetization begins immediately). In terms of policy performance, empirical studies have found that high levels of policy effectiveness are linked to three factors coexisting at the same time:

i. a country's level of policy ambition (for example, level of targets for renewable energy in the overall supply mix),
ii. the availability of well-designed policy incentives, and
iii. the capacity of the system for overcoming noneconomic barriers that may prevent the proper functioning of the market (such as administrative and bureaucratic hurdles and obstacles to grid access) (Miketa & Saadi, 2015; Schwerhoff & Sy, 2017; UNIDO, 2009; Valasai et al., 2017; World Bank, 2018).

H. Qudrat-Ullah, *Understanding the Dynamics of Nuclear Power and the Reduction of CO_2 Emissions*, https://doi.org/10.1007/978-3-031-04341-3_3

The INDC[1]s of the PIS region countries do contain "ambitious" plans and initiatives aimed at meeting the CO_2 reduction targets by the compliance year, 2030. They also mention the need for the development of technical capacity. Both Pakistan and India have some limited success with F[2]IT-based policy instruments (India-INDC, 2018; Pak-INDS, 2016) while Saudi Arabia is in the process of enactment[3] of similar policy instruments and regulations (KSA-INDC, 2015).

In general, five major instruments are often used in combinations to attract investments for renewable and nuclear power technologies. They are:

i. Quantity or Quota-based instruments,
ii. Pricing or Tariffs-based instruments,
iii. Regulations and Standards-instruments,
iv. Public Procurement-based instruments, and
v. Auctions-based instruments.

Most of the prior studies about the effectiveness (or lack thereof) of energy policy instruments emphasize tariff-based policy instruments as an effective practical policy for the development and development of renewable and clean energy technologies (EA, 2020; Lu et al., 2020; Safarzadeh et al., 2020). In this chapter, therefore, we will provide a critical review of these policy instruments together with some successful examples of the implementation of these policy instruments across the world. We hope that this review of policy instruments will support energy policymakers of the PIS region countries and other jurisdictions in their design and improvement efforts for their energy policies.

3.2 Quantity or Quota-Based Instruments

These are the instruments that define a specific target or absolute quantity for renewable energy production in the energy supply mix of the firm or utility or IPP. There are two types of these instruments: (1) Renewable Portfolio Standards (RPSs), (2) Renewable Energy Certificates (RECs). The RPS-based energy policies provide an explicit target for renewable energy (e.g., 30% renewable energy by 2025 or 10% Solar by 2015). The RPS-type policies are aimed at increasing the contribution of renewable energy in the electricity supply mix. The RPS is generally intended to create a stable and predictable market for renewable electricity that maximizes the benefits of renewable generation while minimizing costs (Hamrin et al., 2006). For instance, as a result of effective RP implementations, Australia, ranked 2nd and 5th, respectively, in terms of net capacity additions in renewable power and fuels (not including hydropower over 50 MW) in 2018 (REN, 2021).

[1] INDCs: Intended Nationally Determined Contributions.

[2] FIT: Feed-IN-Tariff.

[3] As an example of policy instrument, UK introduced FIT regulation as a renewable electricity generation support scheme for generators with capacity of less than 5 MW offering a fixed payment per kWh depending on size and type of technology (EA, 2020).

The RPS typically requires that a certain percentage of an electricity supplier or utility's overall or new generating capacity or energy sales must be derived from renewable resources. The RPS features vary from country to country, depending upon their local needs, regulations, and environment, specifically concerning (Hamrin et al., 2006):

- target level,
- whether the target is based on a percent of energy sold or installed
- capacity,
- when targets must be met,
- resource eligibility,
- scope of geographic eligibility,
- preferential policies to encourage particular types of renewable energy, such as specific resource targets or multipliers,
- limits on costs or cost recovery,
- penalties for non-compliance, and
- whether RECS can be used, and/or whether power must be purchased.

The RECs-based instruments are a tradable commodity that represents proof that one megawatt-hour (MWh) of electricity was generated from a renewable energy resource. The RECs can be bought and sold bundled with electricity or unbundled. The RECs are often implemented in combination with the RPSs. The RECs are a highly context-oriented instrument. Three factors determine the success and value of REC (Landry, 2017): (i) demand–supply dynamics (e.g., in the case of Texas, USS, where huge wind generation supply exists, the demand for RECs is depressed), (ii) regulatory dynamics (i.e., jurisdictions with effective RPS implications offer a good chance for RECs to be prized more), and (iii) dynamics of technology (i.e., peoples' perceptions about the utility of various renewable energy-generating technologies differ and they might prefer RECs to promote a specific technology for cleaner electricity generation).

The RPS, on the other hand, can be designed in various ways to achieve renewable energy supply targets. There are some key practices or best practices that policy-makers can avail to design their RPSs. Here are those best practices (Hamrin et al., 2006):

> ***"The purchase obligations should drive the development of new renewable generation.***
>
> *In most cases, RPS legislation intends to increase new renewable energy development. A primary objective of the RPS should be the creation of a stable and predictable marketplace that will support this growth. This can be accomplished by specifying as part of the eligibility requirements that facilities must have become operational after a certain date. Existing renewables can be included as a "baseline" that is built upon through new renewable purchases. Including existing renewable energy facilities in the baseline of an RPS provides financial support for those facilities, helping to extend their operational life. But an RPS that can be satisfied using exclusively existing sources of renewables, such as the RPS in Maine, will not result in any of the goals typically articulated in an RPS law (such as diversifying the energy portfolio, improving the environment, stabilizing fuels costs, meeting load growth, etc.).*

Resource eligibility decisions should be made with care and consistent with goals.

The development of a precise and comprehensive definition of eligible renewables should be based on careful examination of the regionally available resources, the goals of the program, and resource costs. It is also important that supply and demand be balanced with available resources and that those resources can be developed at a reasonable cost. In general, unless specified, the least-cost renewables will be used to meet the mandate. If the development of a specific type of resource is desired, such resources can be differentially encouraged through the use of multiple "resource tiers" or "credit multipliers" (see below).

Purchase obligations should be durable and increase gradually with time.

Increasing the purchase obligation over time will encourage continued investment in new renewable generation and result in a stable, predictable market. For example, an RPS that has a goal of 20% by the year 2015 can have interim targets for minimal annual increases to encourage smooth growth and to avoid 'boom and bust periods. Alternatively, a capacity-based standard might require 1000 MW of new capacity by 2010 with a minimum of 200 MW to be added each year. Overbuilding in one year could be credited toward the following year to encourage economies of scale. This type of ramping-up requirement encourages the construction of the new generation as soon as the standard is put into place while supporting a gradual and continued growth over time. Allowing adequate time for final obligations to be met creates a stable marketplace for developers that also encourages the use of long-term contracts. Because of the substantial capital costs involved in financing renewable projects, long-term power purchase agreements are essential for cost-effectively developing and building renewable generation capacity.

Purchase obligations should be placed equally on all retail electricity sellers, generators, or developers.

Obligations under an RPS generally fall on retailers, but sometimes on generators or developers. Fairness and consistency are essential to ensure that all those who benefit from the increases in renewable supply share in the costs and customers cannot avoid those costs by changing suppliers, as well as costs helping establish a more predictable and stable market for the development of new generation. Connecticut originally exempted utilities from the RPS, but competitive electricity suppliers were subject to the requirement, making it difficult for suppliers to compete on rates with incumbent utilities. Equality in costs helps establish a more predictable and stable market for the development of a new generation.

Programs must have strong and effective enforcement.

Legislation should clearly state what the enforcement policy is and authorize an impartial agency to enforce it. Enforcement is generally the responsibility of the Public Utilities Commission that regulates retail electricity in the state. If there is no signal that non-compliance will result in strong enforcement, RPS targets are likely to be ignored.

Penalties for non-compliance should be higher than the cost of compliance.

Penalties should be effective in creating an environment in which compliance is the best and least-cost option. Financially, compliance with the RPS should be the best outcome for the jurisdictional company. Therefore, non-compliance should result in a financial penalty higher than the cost of complying with the standard. Because the cost of compliance is based on the MW or MWh obligation, using the amount of the shortfall to calculate penalties is a simple way to encourage full compliance. For example, using RECs for compliance can cost a company as little as $2/MWh, while the penalty could be set at $25/MWh, which would effectively set a REC ceiling price of $25/MWh. In some cases, renewable energy from specific technologies can cost more. Setting a specific penalty above the costs of these resources is important if the goal of the RPS is to encourage those technologies. Some states have a specific solar PV requirement in the RPS; solar PV RECs tend to fetch a high *price because their cost is high and supply is scarce. For example, in New Jersey the Solar Alternative Compliance Payment was set at $300/MWh, leading to solar REC prices around $200/MWh. This may seem unreasonable, but it must be financially superior to comply with the standard. In cases with specific resource requirements, penalties should also be specific to these requirements.*

Monitoring "new" requirements for compliance.

In terms of monitoring compliance with a "new" renewable requirement, for a capacity-based standard, this is a very simple process because the mandate is generally for new generation capacity built during a specified compliance period. For a percentage-based standard, generation information systems and RECs can be used to track compliance, allowing regulators to know not only where and when the electricity was produced, but also to match production with specific information about the facility such as the on-line date, re-power date, receipt of tax incentives, etc. The tracking system employed for tracking compliance must be designed in a way that allows for the tracking of this type of "static" information". (Hamrin et al., 2006, pp. 35–37)

Overall, The RPS policies have been popular in recent years with several states in the US and several countries of Europe. The PIS region countries can tap into the potential of the RPS where international donors and investors can be brought online for renewable energy projects both large and small size of power generation facilities. They need to have a sound evaluation of the potentially large investments and the influx of capital in the context of their local conditions (Uddin et al., 2021; Ziras et al., 2021). Local resource use and conservation, preservation of local values and cultural traditions (to avoid any uprising), existing regulations and policies, intramural constraints, financial environment, and adaptability potential of local electricity markets are the critical factors that should inform the decision to the approval of RPSs, especially the foreign-funded ones. Moreover, policy instruments should be approached on a portfolio-basis mechanism—in conjunction with other instruments like FITs together with comprehensive consideration to local conditions that can lead to the effective design of the RPS. Consequently, it can propel the deployment of renewable and clean energy solutions across all of the PIS region and beyond.

3.3 Pricing or Tariffs-Based Instruments

Primarily these instruments are used to reduce financial barriers related to cost and pricing by making favorable pricing regimes for renewable energy relative to other sources of energy supply (Uddin et al., 2021; WB, 2008; Zhao et al., 2013). These instruments are often implemented in the form of (i) fiscal incentives and (ii) Feed-in Tariffs (FIT). Fiscal incentives examples are (i) production/investment tax credits, (ii) capital subsidy, grant or rebate, (iii) public investment, loan or grants, (iv) increase in taxes on competing fuels (e.g., a carbon tax[4] on fossil fuels), and (v) reduction

[4] Economy-wide carbon taxes can achieve net GHG reductions of 13–29% relative to emissions under current policy in 2030, which is equivalent to 27–46% reductions from 2005 levels in the US (Larsen et al., 2018).

in sales, energy, VAT, or other taxes. A range of other supplementary measures is in use that directly stimulates investments in renewable energy, including fiscal and financial incentives, and voluntary measures. A significant and growing category of policies that indirectly promote the development of renewable energy is known as cap-and-trade programs. Austin Energy (Texas, USA) has the most successful Green Pricing in the US, supporting more new renewable energy than any other program in the country (Hamrin et al., 2006). According to Hamrin et al. (2006),

> *One of the reasons Austin Energy has had a successful green pricing program is because of its innovative and unique rate structure. Participants in the Green Choice program see the electric bill standard fuel charge (currently 2.80 cents per kWh, but is subject to fuel adjustment) replaced by a Green Choice charge of 3.30 cents per kWh of electricity used. This means that customers typically pay about one-half cents more per kWh to help support the renewable energy power provided by Green Choice. The flat green rate provides customers with a price hedge against volatile fossil fuel prices; the Green Choice rate is a fixed rate. While fossil fuel prices are unstable, their product is offered at a fixed rate. Green Choice's largest resource consists of new wind turbines in West Texas. The program also receives electricity from four new landfill methane gas projects located around Texas. Austin Energy has signed 10-year contracts for electricity from the wind and methane gas projects outlined above. The price for that electricity will remain the same for the life of those contracts, allowing Green Choice customers a way to hedge against fossil fuel price volatility.*
>
> *The Green Choice program was authorized by the Austin city council in 1999 and the program was launched in 2000. The initial rate was set at 1.7 cents per kilowatt-hour. This rate did fully recover the costs of the original green power sources and was subsidized up to $1 million. Ten months after launching its program, Austin Energy had fully subscribed to its initial 40 MW of new renewable supply and had to contract for additional renewable supply. Austin's second offering was not subsidized and was priced at 2.85 cents per kWh. This represented the contract price for the wind and did not include congestion or ancillary services costs, which were not anticipated. At the end of 2003, Austin Energy increased the green power rate to 3.3 cents per kilowatt-hour. This new rate covered the wind contract price, congestion costs, and ancillary services costs. The new rates apply to new program subscribers only; existing subscribers continued to pay the lower green power rates established in earlier phases of the program. At the same time, standard fuel charge rates were changing as well. This made the difference between standard service and green pricing larger or smaller. The price difference between standard electric service and Green Choice will fluctuate over time because of changes to the fossil fuel charge. Only Green Choice customers get the benefit of ten years of price stability, and that level of price certainty is very attractive to customers, particularly large consumers of electricity. Currently, the price of their renewable energy product is lowered than the price of their default service, creating a "negative premium" for green power customers.* (Hamrin et al., 2006, pp. 64–65)

Most of the FIT contracts will have three key tenets in them: (i) guaranteed grid access, (ii) guaranteed power purchase agreements, and (iii) long-term preferential rates for the clients' excess electricity (Qudrat-Ullah, 2015; Uddin et al., 2021). One might ask what is the relevance of this example from the US to the countries of the PIS region. Well, here it is: (i) empirical studies have shown that it is feasible and essential for the energy policymakers of developing countries to learn the best policy practice from successful countries to promote their renewable and cleaner electricity-generating technologies development and deployment in their countries (Zhao et al., 2013), and (ii) the electricity tariffs paid by the consumers are rising faster than any increase in their income, so the affordable electricity aspect of the stated goal of the UN's SDG No. 7 is likely to be compromised in these countries. Therefore, it is hoped that countries of the PIS region can at least act similarly to the experience of those countries with a successful implementation of various policy instruments and energize their countries with *reliable, affordable, and cleaner electricity* with the deployment of renewable and clean electricity-generating energy technologies.

3.3.1 FIT Payment Models

FIT-based energy policies have to have three provisions: (i) some kind of preferential tariff or a premium price to be paid to the power generators/suppliers), (ii) guaranteed access to the grid, (iii) guaranteed purchase of electricity produced by the FIT program participants/producers for a specified period (often comes as a power purchase agreements (PPAs). The first four countries to introduce some type of preferential tariff or FIT were India (from 1993), Sri Lanka (from 1997), and Brazil and Indonesia (from 2002) (REN21, 2016).

FIT policies have various payment models. Because each renewable energy generation project is unique, differentiation of FIT payments to account for these differences can ensure that a variety of renewable technologies and project sizes are realized (Cory et al., 2009). One primary FIT payment model is whether the project owner's compensation is tied to fluctuations in the actual market price of electricity. These two different policy options are often characterized as either fixed-price or premium-price policies. These two models dominate FIT policy design; however, most countries with FIT policies choose the fixed-price approach (Klein et al., 2008; Mendonca, 2007). Figure 3.1 illustrates a fixed-price FIT model. In this policy design, the total FIT payment to the project remains independent from the market price and is a predetermined payment for a guaranteed period. Because fixed-price FIT policies offer market-independent payments, they create stable conditions for investors.

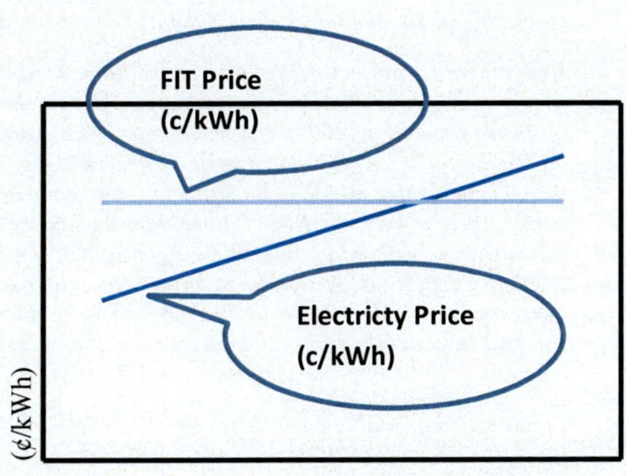

Fig. 3.1 Fixed-price FIT model (*Source* Adopted from Cory et al., 2009)

This risk reduction can lead to lower project-financing costs (de Jager & Rathmann, 2008), making renewable energy even more competitive.

According to Cory et al. (2009),

> Another important distinction in feed-in tariff design is how the payment levels will be differentiated, based on project-specific factors. These factors can include the technology type (whether solar, wind, geothermal, etc.; or the fuel type, in the case of biomass and biogas), the size of the project (to account for economies of scale), the quality of the resource at that particular site (to encourage broad deployment of wind and solar power, and limit windfall profits at high-quality sites), and/or the specific location of the project (e.g., building integrated, offshore wind). (Klein et al., 2008)

3.3.2 Successful Factors for FIT-Based Policies

FIT-based policies to promote renewable energy need support mechanisms to succeed. Researchers have identified six factors for the success of FIT policies to attract renewable energy projects: (i) Project-financing support, (ii) Cost-effective procurement mechanism), (iii) Hedge against project delays and cancellations, (iv) Focus on "reasonable-cost" renewables, (v) Assured support for emerging technologies, and (vi) Provide ratepayer backing. According to Cory et al. (2009),

> ***"Project-financing support.*** *Not all states have RPS design elements that support new project financings, such as a requirement for long-term contracts or centralized state procurement (*Wiser & Barbose, 2008*). Without long-term support to secure investment, renewable projects will likely have difficulty securing financing (Cory et al.* 2004*), which could result in a shortage of supply to meet RPS demand. FIT policies provide the revenues that project investors require and can ensure that enough supply will come online.*
>
> ***Cost-effective procurement mechanism.*** *Due to the guaranteed contract terms and the stable investment environment created by FITs, these policies appear to be a cost-effective procurement mechanism for renewable energy development. They could be used alongside competitive solicitations; or, provided the FIT payments are differentiated to account for economies of scale, they could be used to replace competitive solicitations to meet government-established renewable goals, similar to what is done in countries such as Germany and Spain.*
>
> ***Hedge against project delays and cancellations.*** *Among other things, project siting and access to the transmission can challenge even the best and most economical renewable projects (Wiser et al. 2005). If a utility's renewable procurement process does not consider the likelihood that a project will be developed (and just looks at the lowest cost, for instance), then it is likely that not all of the projects under contract will be built – the utility, therefore, is less likely to meet its RPS. Rather than having the utility determine which projects go forward (i.e., with whom it will sign contracts), the government or utility can establish eligibility criteria as well as a payment level under a FIT – anyone who qualifies and is interested in investing in RE technology can do so and obtain a standardized utility supply contract (without the transaction costs or any potential gaming). This can help ensure that the best portfolio of projects moves forward.*
>
> ***Focus on "reasonable-cost" renewables.*** *Similar to other power production, utilities must justify their costs for RPS compliance, whether through power purchase agreements or utility-owned projects. While the focus on "least-cost" principles attempts to minimize ratepayer costs, they may pressure utilities to negotiate contract prices for renewable projects that are*

inadequate to secure financing (and fail to adequately address investor risks). Instead of focusing on the least cost, FIT policies focus on estimates of the actual costs required to build renewable projects based on technology and other project-specific considerations. If designed well, the FIT can ensure that a variety of projects receive just enough to cover their costs and a reasonable return.

Assured support for emerging technologies. *New or emerging technologies may not be able to secure financing, even with long-term utility contracts. The projected revenues need to be high enough to support the additional investment risk faced by investors. This higher risk requires higher-equity returns than commercially available renewable energy projects. Appropriately, structured FIT policies will include this risk premium for emerging technologies (paid for by the ratepayers) and provide the long-term assurance that investors require.*

Provide ratepayer backing. *Regulated utility generation is sometimes subject to "prudence" reviews of investments and contracts after projects are built. If costs are deemed to not be prudent, the utility will have to cover the costs itself instead of relying on ratepayers, sometimes retroactively. Ultimately, this means that utilities may be uncertain as to whether they will be able to recover the costs from a contract or the ownership of new renewable projects. Overall, the FIT structure can provide more certainty, because the FIT payments are backed by the ratepayers and typically are not subject to retroactive regulatory prudency review. This certainty can help utilities become interested in FIT policies, particularly if the utilities are eligible to participate as project owners*". (Cory et al., 2009)

For the PIS region, the Indian experience with FIT policies is substantial. According to WNA, "In [the case of] India, ten out of 29 states have feed-in tariffs, eg 2.75 times the tariff for coal-generated power in Karnataka, plus a federal incentive scheme paying one-third of the coal-fired tariff" (WNA, 2021).

Overall, FIT policies have proved successful in bringing new renewable energy in the world. There is no one FIT model (say for tariffs, payment modes, terms of the contracts, etc.) that could be used as it is for a country in the PIS region. For instance, FIT tariffs can be technology-neutral vs. specific technology-based, flat vs. stepped tariffs, fixed vs. premium tariffs, and constant over time vs. declining over time. In contrast to other financial incentives for renewables, FIT-based tariffs do not decrease a developer's up-front costs. To overcome these upfront high capital costs, policymakers can enact investment tax credits, grants, and rebates. Moreover, grants and rebates can be integral in increasing the market penetration of small, customer-sited projects. For instance, to lower the burden of added tariffs on poor nations, Global Energy Transfer (GET) FIT is presented as the proposed solution to leverage international Public–Private Partnerships to support and de-risk national FITs in Uganda (GETFIT, 2016). Its key purpose is to mobilize capital by providing financiers transparency, longevity, and certainty.

Which model of FIT is good for the countries of the PIS region? Depending on its objectives and national aspirations, each country can apply a tailor-made FIT-based renewable energy policy to attract and promote renewable energy investments. This customized approach to the design of the FIT policy for renewable energy is particularly in congruence with the electrification programs of the PIS region countries, especially for Pakistan and India where substantial rural areas are still either without any access to electricity or face severe shortages on an ongoing basis.

3.4 Regulations and Standards-Based Instruments

The policy instruments under this category are mandatory as dictated by the regulatory and legislative body of a country. Regulatory standards can promote renewable energy both directly and indirectly. For example, California enacted its regulations in 2018, which require the state's utilities to generate 100% clean electricity by 2045. Several countries have mandated utilities to deploy net metering. Net metering allows a two-way flow of electricity and only charges consumers for their net electricity use.. A common goal of these regulations, standards, and instrument is to supply electricity from the consumer to the grid at times when generation exceeds consumption (e.g., excess electricity might be generated by a solar-PV system during the day and consumer-generator will need it during the night when the sun has already set) and returning it in the opposite case. For the successful effects of these renewable and clean energy promoting incentives and mechanisms like "net metering", the transparency of policy instruments is instrumental (Ziras et al., 2021).

In a direct form, a country's policy objectives are set to favor renewable energy over non-renewable fossil-based power generation by removing non-financial barriers. By imposing restrictions on the fossils-based generation and increased carbon limits and taxes, these regulations and standards can indirectly support the deployment of renewable energy technologies in-country (Benitez, 2012). For example, on September 24, 2004, after a year and a half of public hearings and participation by over 150 parties, the New York State Public Service Commission (PSC) issued its "Order Approving Renewable Portfolio Standard Policy," which is the PSC's renewable energy policy and provided definitions and targets for carrying out the policy (Hamrin et al., 2006).

According to Hamrin et al. (2006) findings, the best practices for regulations and standards are:

- Regulators should require periodic reporting of program performance by utilities and suppliers to ensure that cost recovery is warranted.
- Utilities could be given incentives to enhance the success of check-off programs.
- Limit the number of participating suppliers to three, at most to have a successful, easy-to-administer program.
- Select the supplier(s) through a competitive process to obtain the least-cost resources.
- To ensure that the programs result in additional renewables that would not otherwise have been constructed, suppliers should provide products that are based on new renewable resources, preferably covering a substantial portion of a customer's electricity usage.
- Provide a mechanism for utilities to recover costs for their administrative and marketing support of the program. If utility costs are not covered, they will not be motivated to promote the program or even get it off the ground (Hamrin et al., 2006, p. 19).

These best practices in terms of the design of regulation and standards-based policy incentives should guide the concerned regulatory bodies of each of the PIS region countries to chart its course of actions and regulations that will accelerate the availability of cleaner electricity in their countries. In the case of Pakistan, for example, "net metering" can provide several befits to the major stakeholders in distributed generation of electricity including policymakers, distributions companies (i.e., DISCOs[5]), and the consumers. Policymakers can promote renewable and clean energy with minimal need of new regulations, DISCOs can benefit with (i) additional capacity, and (ii) lower T&D losses, and the consumers can lower their bills and enhance their savings (NTDC, 2018). With such a beneficial promising potential, the government of Pakistan should step up its efforts to implement this clean-energy supplying regulatory and standard-based mechanisms. Although some forms (e.g., commercial, residential, and industrial uses) of "net metering" are made available now, however, the availability of net metering and risk-reducing initiatives by AEDB and others need to be publicized through a targeted consumer awareness campaign (USAID, 2019).

3.5 Public Procurement-Based Instruments

Governments are often a very large energy consumer in the world. Specifically, in the PIS region where the commercial and industrial sectors developing but at various speeds, the residential sector is the often major consumer of energy. Through large public procurement, a government can affect the energy market and can use procurement requirements as a tool for national and regional governments to develop renewable-energy markets and build capacity (Benitez, 2012). However, to fully avail the benefits of these public-purchasing based instruments, impeding factors like (i) lack of ownership by purchasers, (ii) failure to address the whole cycle acquisition, and (ii) poor management of risk aversion, should be resolved effectively (Georghiou et al., 2014). Attracting investments for the promotion and deployment of renewable and clean energy technologies requires concerted efforts on the part of the energy policymakers to institutionalize the best practices and procedures regarding public procurement.

Sustainable public procurement is a tool that provides a significant lever for governments to accelerate the shift towards more sustainable consumption and production patterns, and more generally to contribute to the achievement of sustainable development goals (Kjöllerström, 2008). Governments at various levels (e.g., city, province, and federal levels) in a country can enforce regulations for all kinds of procurement to be green, enabling renewable energy technologies deployment in its country. The IEA has strongly recommended that governments should stimulate investment using public procurement in markets such as carbon capture, usage and storage (CCUS), CO_2 utilization, and low-carbon products, following the lead of the Netherlands, Canada, and India (Kjöllerström, 2008). In the case of Canada,

[5] DISCOs: Distributions [of electricity] Companies [of the government of Pakistan] (NTDC, 2018).

for example, governments at each level ask their supplier to comply with low-carbon footprint standards, indirectly favoring renewable energy-based production and consumption.

Another tool that governments use to achieve specific objectives in the context of efficiency and renewable technology deployment is "public benefits fund (PBF) (Nadal & Kushler, 2000). A public benefits fund (PBF, or "fund") is a revenue stream most commonly financed through an ongoing surcharge on consumer electric bills (e.g., a "green tariff"), but also occasionally established through lump-sum cash transfers required by state legislation or regulatory settlements. It is used to directly support projects and activities in the electricity sector that provide important public benefits or overcome market barriers. Roughly half the states have established PBFs to promote investments in energy efficiency and/or renewable energy technologies (Hamrin et al., 2006).

The best practices for the establishment and use of PBF are:

*"**PBF Creation***

Absent other opportunities (e.g., utility merger or environmental settlements), finance the PBF through a volumetric charge on electric bills in a way that is competitively neutral and non-bypassable (i.e., consumers cannot avoid paying the charge by switching electric suppliers).

- *Set the funding level as high as is politically feasible. At current funding levels in the United States, public benefits likely far outweigh the costs. Under-funding limits the programmatic opportunities available to a program.*
- *Provide long-term funding stability. Because it takes time to implement programs effectively and build markets, a minimum of five to seven years of funding stability should be provided. Shorter funding cycles could preclude risk-taking and the implementation of multi-year programs aimed at the market transformation.*
- *Insulate the fund from budgetary raids. Involving a wide variety of stakeholders in the creation of the fund could help to build political support. A dedicated source of funds (e.g., an electricity surcharge) might be less vulnerable than funding from more indiscriminate governmental sources of money (e.g., settlement funds). Minimizing carryover of funds from one year to the next may make for a less-tempting target.*
- *Finally, the legislative language that authorizes the use of funds only for specific purposes (as in California) can be helpful but still may not prevent the government from "borrowing" the funds under an indeterminate repayment schedule or altering the legislation to allow a broader re-appropriation of funds.*

Programmatic Scope

- *Clearly define which renewable resources and technologies are eligible for PBF funding. The use of general terms such as "biomass" or "customer-sited" can, without further clarification, create funding uncertainties, and potentially generate legal action, that could disrupt or delay the development of entire industries.*
- *Provide clear guidance for the allocation of PBF funds across resource or technology groups to avoid fighting among different renewable energy industries (e.g., solar vs. wind). At the same time, PBF administrators should retain sufficient flexibility to shift any prescribed allocation of funds in response to changing market conditions or emerging opportunities.*

- *Geographic restrictions (if any) on where funded projects can be located should be defined within the context of in-state renewable resource availability as well as the structure of the electricity market. For example, if the electricity market is regional (e.g., as in New England), then funding for out-of-state projects within the region might be acceptable. PBPs can require that power and/or RECs from funded out-of-state projects be delivered into the state to ensure that at least some of the project's benefits accrue to in-state ratepayers.*
- *Allow the fund administrator sufficient flexibility to choose the types of programs it will offer, as well as the types of funding recipients it will target. For example, one PBF in the United States is only able to offer consumer loans, while another is only able to fund non-profit organizations; both have found these severe restrictions to be detrimental to their overall mission.*
- *Set clear and reasonable goals from the start, but also allow the fund to shape the measurement and evaluation process to reflect the types of programs that it offers. Early and clear demonstration of program effectiveness and success through independent evaluation may help protect PBFs from budgetary raids.*
- *Seek out and encourage regional coordination with other states' PBFs. Most funds, particularly those within the same region, share common needs and experiences and could learn from other funds, as well as benefit from the potential economies of scale that regional coordination might provide. In the United States, the Clean Energy States Alliance serves as a conduit for information-sharing and joint project activity among renewable PBFs.*

PBF Administration

- *Carefully weigh the pros and cons of different administrative options. The presence of adequate, experienced, capable and dedicated staffing is likely to be a more important determinant of success than whether the fund is administered by a utility, a government agency, or a new or existing non-profit organization.*
- *Identify a single entity charged with oversight of the PBF administrator and clearly define the extent of oversight responsibility and authority. Without a clear "chain of command," PBF administrators may feel compelled to answer to multiple interests (e.g., the PUC, the legislature, the governor), thereby needlessly adding to the administrative burden.*
- *To avoid under-staffing, designate (and perhaps set limits on) explicit funding for administrative costs. On a percentage basis, it is not uncommon for 5–10% of PBF funds to be used to cover administrative and management costs. Some funds have set limits on the proportion of the fund that can be used for such costs. Xcel's Renewable Development Fund has a 5% cap in place, NYSERDA has a 7–8% cap, and Oregon caps administrative costs at 11% of funds (but currently only spends about half that much)."* (Hamrin et al., 2006, pp. 43–45)

The main purpose of these best practices to the various aspects of a PBF is to provide the policymakers a benchmark to compare their practices or begin an afresh journey on this track to facilitate the development of renewable production and procurement in their countries. Learning from others' experiences will allow the energy policymakers of the PIS region countries not to make costly mistakes when it comes to the design of policy incentives for the deployment of renewable technologies and energy supply in their own countries. As a result, one could expect a boost for much-needed cleaner electricity in each country of the PIS region.

In the case of Pakistan and India, both are heavily fossil-import-dependent and Saudi Arabia is the largest consumer of domestic oil and gas production. By enacting and enforcing regulations regarding its procurement activities, it can facilitate the development and adoption of renewable power-based production. By facilitating the deployment of renewable technologies across the sectors of its economy, a country can expect a substantial increase in renewable power generation in its country. To meet the requirements, producers might well adapt to renewable energy at the source that in turn can help the renewable and clean electrification programs (e.g., through the establishment of small green production centers in rural areas). Associated non-fiscal incentives (e.g., preferred supplier status to those who are 100% in compliance with green and renewable energy-based production and consumption) can further support the development of the market and capacity for renewable and clean energy in the country.

3.6 Auctions-Based Instruments

An auction is a selection process to allocate goods and services competitively, based on a financial offer—e.g., you place an item at eBay and people will bid and the highest bidder gets the deal. In a 'reverse auction', electricity generators bid their supply to distribution companies in the electricity market of a country or region, and the process is designed to select the lowest price. Auctions can be used to discover appropriate tariff rates for FIT policy and auctions can be a very attractive mechanism for attracting new renewable energy supply in a country (Benitez, 2012). In the case of the deployment of renewable electricity, auctions, primarily due to their efficiency, have become the main instrument of choice to support renewable electricity around the world (del Río & Kiefer, 2021). They, utilizing a database of 90 renewable electricity auctions from around the world, found large differences across regions and over time regarding the use of auctions for the promotion and deployment of renewable energy. Thus, these auctions should be designed based on the local condition of the host country.

For large-scale on-grid auctions, the host utility should create sound PPA on a long-term basis for the success of auctioned projects of renewable energy, Auctions bring transparency to renewable energy projects and alignment of industry and government knowledge about the development and operations of renewable energy technologies. In auctions, policymakers can manage both the price and the quantity of renewable energy produced by providing revenue guarantees for project developers (similar to FIT) while ensuring that renewable generation targets are met (Wyman, 2017). Actions also governments and policymakers customize these actions to their needs and objectives of the renewable energy policy. Once projects are assigned, for timely completion of these renewable energy projects the establishment and implementation of enforcement regulation are critical.

According to Oliver Wyman (2017), the key success factors for renewable energy auctions (in the context of mini-grid-based renewable energy supply) are:

- Clear auction terms and conditions that are aligned to detailed and consistent regulations1 which enable successful auction processes and project implementation,
- Attracting multiple bidders to ensure competitive tension and desired capacity is met by suitable developers,
- Having a fair competition in which multiple bidders compete in a non-collusive manner in a transparent auction,
- Ensuring that the project is successfully delivered on time after the auction process,
- Minimizing the auction costs for all stakeholders,
- Achieving the lowest-priced bids or lowest possible subsidy whilst successfully fulfilling other criteria, and
- Meeting any developmental goals that have been set. (Wyman, 2017, p. 14)

When policymakers want to design auctions for renewable energy capacity expansion, there have to consider certain trade-offs among the different design elements including (i) auction's demand, (ii) bidders' qualification requirements, (iii) winner selection process, and (iv) seller's liabilities. IRENA and CEM (2015) have provided guidelines about these design elements of auctions. These guidelines are:

*"**Account for trade-offs between different design elements***

*When selecting design elements, policymakers should carefully consider the inherent trade-offs between potentially the most cost-effective outcome and other objectives. In defining the **auction's demand**, ambition for a greater role of renewables in the energy mix must be weighed against cost-effectiveness.*

- *When the objective is to develop a specific technology, policymakers may want to select a technology-specific auction – one of the ways of defining "exclusive demand bands". If the goal is minimizing costs, a technology-neutral auction can be introduced, allowing competition between technologies, therefore favoring the more mature and cost-competitive technologies.*
- *When the objective is to meet urgent capacity needs while retaining flexibility in holding auctions, policymakers may auction the total volume at once through a standalone auction. If the objective is to further enhance investors' confidence for a most cost-effective outcome, the total volume auctioned, if considerable, can be divided into different rounds in a systematic auctioning scheme, with a set cap on the volume auctioned in each round. This facilitates long-term planning by policymakers, bidders, and renewable energy equipment suppliers, which may be beneficial to the country's renewable energy industry and the grid extension planning.*

*In establishing **the qualification requirements**, there is a trade-off between reducing entry barriers to encourage competition and discouraging underbuilding.*

- *Allowing the participation of a large number of bidders while ensuring that they can successfully deliver the project requires a careful selection of qualification requirements. While the requirement for an extensive track record in the field, for example, can help ensure timely project completion, it may also limit the participation of new and/or small players.*
- *Specific renewable energy deployment goals can be reached through qualification requirements, such as technical requirements, project size requirements, or location constraints. Although they can lead to desirable outcomes, they may increase the contracted price, as developers need to adapt their projects to these requirements.*

- *If the objective is to also meet broader development goals, policymakers can include additional selection criteria. Local content requirements, for example, can support the local industry, job creation, and other socio-economic benefits. Such requirements are most effective when aligned with other design elements, such as a long-term auction schedule, and applied with other supporting policies.*

While a simple ***winner selection process*** *provides greater transparency, some degree of complexity may have to be implemented to ensure that the objectives of the country are achieved by the auction.*

- *If the objective is to reach the lowest price using a simple procedure, policymakers can choose to adopt the classical minimum-price criteria for the selection of a winner. However, other objectives can be achieved by incorporating non-monetary criteria in the process, such as socio-economic benefits, location, developer's experience, etc. This may, however, result in higher prices and a more complex mechanism.*
- *When the main objective is to ensure cost-effectiveness, policymakers can also set a ceiling price above which bids are not considered. However, if the ceiling price is not calibrated properly, there is a risk that a suboptimal amount of renewable energy will be contracted, as it could lead to the outright rejection of certain perfectly reasonable bids. Experience has shown that keeping the price ceiling undisclosed can help increase the cost-effectiveness of the scheme but at the risk of disqualifying potentially good projects that are just above the ceiling. Disclosing the ceiling price in auctions where competition is not fierce, might result in equilibrium prices right below the ceiling.*

In determining ***the sellers' liabilities in the contract****, there are various ways to allocate risks between the project developer, the auctioneer, and the contract offtake including financial, operational, and project implementation risks. The over-allocation of risks to developers impacts the level of participation of bidders and ultimately the contracted price.*

- *To limit the risk of delays and underbidding, policymakers can enforce stringent compliance rules, but at the expense of increasing transaction costs, which in turn may limit the participation of bidders and also result in a price increase.*
- *Developers might be subject to risk, but they should not be subject to uncertainties. The risk allocated should be communicated, transparent, fully quantifiable, and enforced. Protecting possible bidders against uncertainties is key to gaining their confidence.*
- *The auctioneer should ensure that the compliance rules and penalties included in the auction are enforced.*

Ensure transparency to increase developers' confidence

Attracting bidders is key to the success of an auction. Transparency, simplicity and the developers' perception about the fairness of the process increase investors' confidence.

- *The auctioneer must define fair and transparent rules and obligations for all stakeholders. Any information or adjustments about the bid must be communicated to all competitors equally (e.g., a dedicated website, conference at the start of the auction, etc.). Policymakers need to consider evaluating the process at the end of each round, as it is important to factor lessons learned into the design of the following rounds.*
- *Administrative procedures should be simplified, streamlined, and facilitated when possible (permits, grid connections, etc.). Setting up a one-stop-shop could help minimize transaction costs and efforts of the bidders, preventing delays in project implementation. Also, the time, manpower, and skills needed to evaluate bids have to be carefully estimated.*

- *Policymakers should minimize the investors' perceived risk through an institutional and regulatory framework that ensures a predictable and stable environment for investments. Good auction design is not enough in a market in which the level of skepticism is high and the credibility of the auctioneer is in question.*

Tailor the design of auctions to the specific context

- *There is no "one-size-fits-all" formula for successful auctions. Different design elements should be selected and combined in a way that is tailored to meet the goals of the auction, according to the country's specific requirements and characteristics. While determining which auction design best fits the specific context, policymakers should take the following types of constraints into account: those arising from the macro-economic conditions (local and global), the characteristics of the power sector, and the inter-dependencies between design elements.*
- *All the design elements, examples, and other recommendations are analyzed and illustrated in this study on Renewable Energy Auctions: A Guide to Design* (IRENA & CEM, 2015, pp. 38–41).

Overall, these directives and guidelines are a support mechanism for the policymaker and the designers of auction-based policy instruments. In the context of Africa, in South Africa in 2011, wind auction requirement of 25% local content, which the government aims to raise step-by-step to 45% (first bid submission phase), 60% (second phase), and 65% (third phase) and for solar PV, the local content requirement rose from 28.5% under the first round to 47.5% in the second (IRENA & CEM, 2015).

According to Oliver Wyman (2017),

i. In the case of Zambia, experience shows that renewable energy auctions with involvement from World Bank achieved record low prices, auctions can be a cost-efficient way to bring international best prices into new markets, international organizations can play an important role in mitigating political and counterparty risk, and reducing the cost of capital, and thorough diligence for site selection is important;
ii. South African experience shows an example of a dramatic decline in prices due to multiple auction rounds and changing key auction design elements. Auctions can be successful in encouraging a significant scale-up in renewable energy projects. Revealing ceiling prices can lead to an anchoring effect as bidders use a price cap. Government commitment and private partner-friendly policy environment are important; and
iii. If multiple agencies are involved in an electrification program, the responsibilities of each must be clearly defined to prevent friction between them. Grid expansion is a serious threat to developers. Direct competition between a subsidy-backed utility and private companies should be avoided" (Wyman, 2017, p. 16).

The success of a particular auction strongly depends on the given framework and market conditions prevalent in the host country (Haufea & Ehrhart, 2018). In other words, "there is no one model that fits all." For instance, competition increases with a higher number of bidders and by replacing weaker bidders with stronger ones (Kreiss,

2016). Countries in the PIS region can use auctions to develop their local markets for equipment and installation needs with the aim to achieve self-sufficiency over a stipulated period. Renewable energy technology development and deployment can get support from a gradual approach based on actions by the regulatory agencies of the PIS region countries. With cleaner electrification at the core of their energy policy, local capacity development in renewable technologies is only a winning proposition as a complementary benefit of the auction processes.

3.7 Policy Instruments in Actions

Most of the countries of the words have embraced a variant of the above-mentioned policy incentives for renewable energy support. They have based their energy policies on three main categories of instruments: tariff-based, quantity/quota-based, and hybrid (please see Table 3.1). A country's existing infrastructure, existing energy policies, political will, community participation and ownership, electricity market structure are the factors that would shape the type of policy instruments suitable for that country.

3.7.1 Global Experience with Policy Instruments

Figure 3.2 displays the use of various policy instruments for renewable energy by several countries of the world. It is worth noting that almost all countries apply some type of fiscal or financial incentive in parallel to price or quota-based mechanisms. For instance, while India has used FIT, Auctions, and REC/TGC (Tradeable Green Certificates) to promote investments and generation of renewable energy, Turkey on the other hand has relied on FIT policies to grow renewable energy production in the country. Also, it is notable that 28 developing countries have based their renewable energy policy on FIT-based instruments. One can make a resealable assumption that the PIS region nations can also befit from FIT-based policy instruments incorporated into the design of their renewable energy policies. As FIT policies are more attuned to the small-scale deployment of renewable energy generation technologies, it is expected that to advance the cleaner production of electricity in the PIS region., FIT-based policy instruments are most likely to be adopted there.

Table 3.1 Fundamental types of policy instruments categories

	Price-driven	Quantity-driven
Investment focussed	Rebates Tax credits	Auctions
Generation-based	FIT and its variants	Quotas/RPS

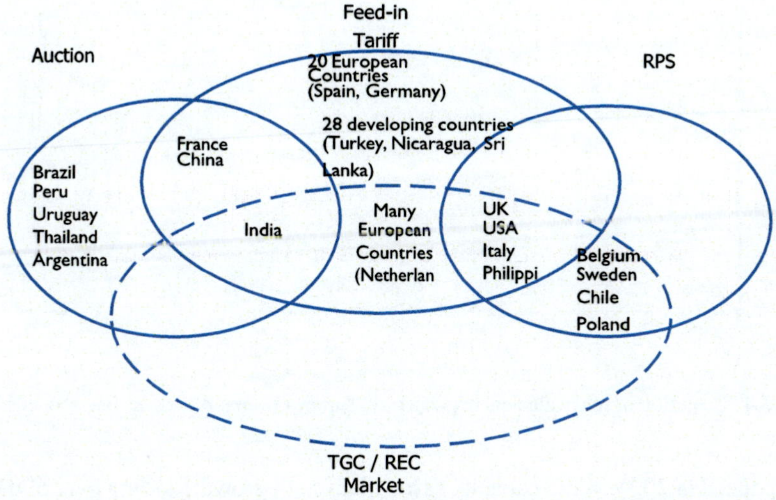

Fig. 3.2 Use of renewable energy policy instruments (*Source* Azuela & Barroso, 2011)

As a result of the implementation of policy instruments, each country has experienced a different trajectory of renewable energy capacity development. For instance, in the case of India, as is shown in Fig. 3.3, post-2002 when the FIT policy was introduced, renewable capacity took off at a relatively faster pace. Wind power being the biggest beneficiary of the FIT policy, wind capacity was more than quadrupled in about seven years. Followed by wind-capacity, biomass-capacity (almost tripled during the same period), and small hydropower also witnessed the growth incapacity (almost doubled during the period from 2002 to 2009).

In the case of Turkey, as can be seen in Fig. 3.4, the Feed-in Tariff policy was

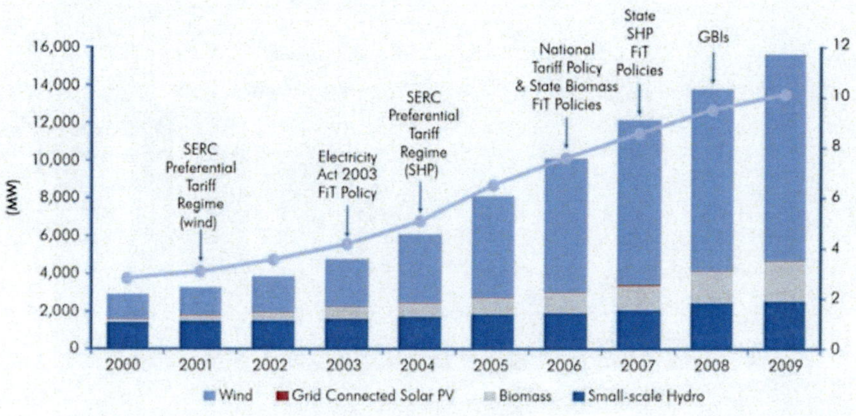

Fig. 3.3 Evolution of renewable derby capacity and share, India (*Sources* EIA, 2009; Ministry of Power India, 2010. *Note* India introduced a REC market and the use of auctions from 2010)

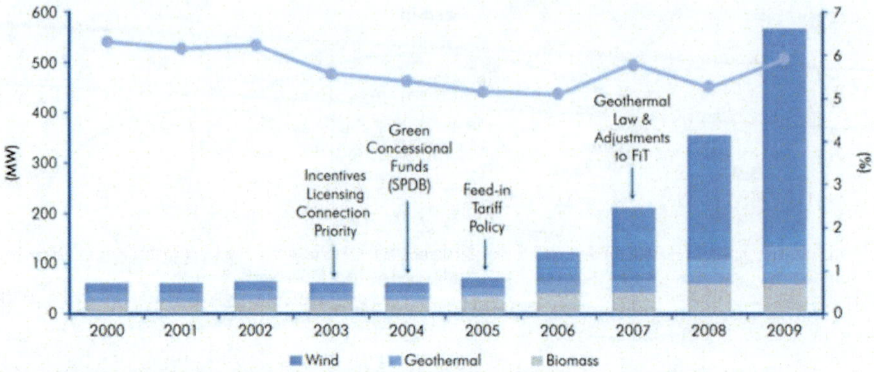

Fig. 3.4 Evolution of renewable energy capacity, Turkey (*Source* Azuela & Barroso, 2011)

introduced in 2005. Wind capacity saw a mammoth growth, adding over 600 MW of capacity in just four years. Geothermal and biomass capacity was almost double in these post-FIT policy four years. Turkey should capitalize on the policies that have brought this success so far.

Figure 3.5 portrays the evolution of renewable energy capacity in Sri Landa, a developing country. This growth of renewable energy capacity pints to a different trajectory based on local resource base specialty. The government of Siri Lanka introduces fiscal incentives and regulations in early 2000 which resulted in the development and growth of small-scale hydropower: the hydro capacity increased from a mere under 20 MW in 2000 to over 120 MW in 2009. Although the government made some adjustments in renewable energy policy incentives in 2006 and 2008, the growth trajectory remained consistent. It appears as if wind capacity remained stagnant through this period.

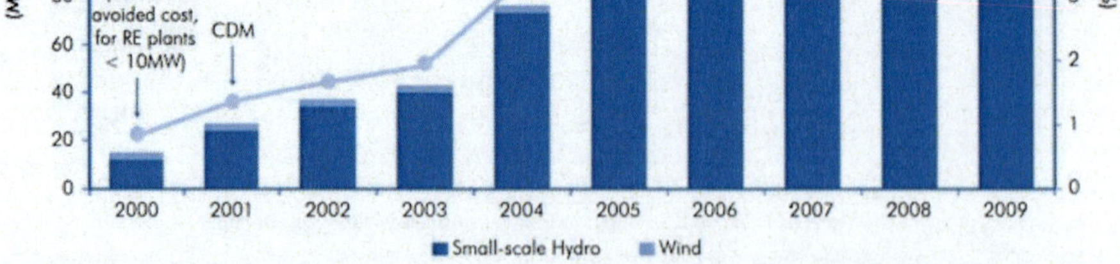

Fig. 3.5 Evolution of renewable energy capacity, Sri Lanka (*Source* Azuela & Barroso, 2011)

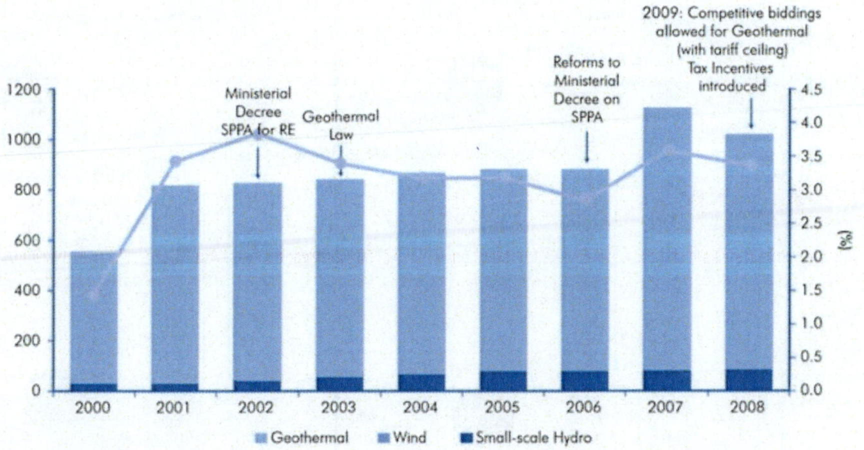

Fig. 3.6 Evolution of renewable energy capacity, Indonesia (*Source* Azuela & Barroso, 2011)

The renewable energy capacity trajectory in the case of Indonesia, another developing nation with a large population. As can be seen in Fig. 3.6, the dominant renewable energy technology is geothermal. Although the Indonesian government has been introducing regulations and incentives (e.g., a Geothermal Law in 2003, the geothermal capacity hardly saw significant growth. The small hydropower did see some growth over the 2000–2008 period. Surprisingly, wind capacity (despite the huge potential capacity) did get much traction. It appears that policymakers need to adjust the policy instruments to allow growth and diversity in the renewable energy capacity of Indonesia.

As we have discussed earlier, the design of policy instruments is something different and the actual development of renewable capacity is quite another task. To realize the targeted capacity, there are various non-financial barriers (e.g., cultural acceptance of new technology, bureaucratic procedures, etc.) that have to be overcome. Figure 3.7 displays the gap between the evolution of average market growth and the required market growth to realize the targeted capacity in these countries. Except for India, other countries on this chart (Turkey Siri Lanka, and Indonesia) have to do a lot before they can that their targets of renewable energy are achieved. India also has to reinforce its policy incentive to reach its target capacity for renewable energy.

To achieve a certain supply mix of renewable energy, governments provide subsidies to support policy instruments' implementation. For example, in the case of India, to achieve a renewable capacity of 45 MW by 2020, the estimated subsidies per annual basis are in the range of 1.2 Billion US dollars to 1.8 Billion US dollars (ESMAP, 2010; Ministry of Power India, 2010). On the other hand, to achieve a diversified mix capacity of about 1200 MW of renewable energy in Siri Landa, the government has to pay annual subsidies in the range of 17.2 Million US dollars to 23.2 Million US dollars (Azuela & Barroso, 2011). These subsidies can be seen as

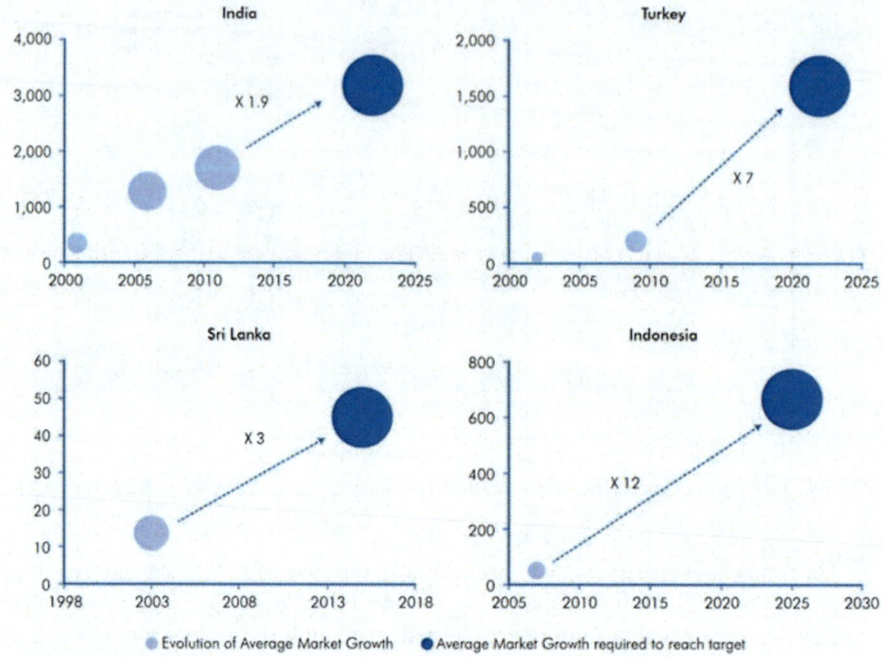

Fig. 3.7 Evolution of reaching targets for renewable energy (*Source* Azuela & Barroso, 2011)

long-term investments for the development and deployment of markets and technologies for renewable energy. However, each country should have a sound plan and energy policy towards the gradual withdrawal of these subsidies over time.

Overall, when it comes to the design of policy incentives, there is no specific relationship between the type of policy incentive (e.g., FIT or RPS) and the market structure (more liberalized (e..g, Turkey) versus highly centralized electricity markets. Also, the most effective policy incentive may not be the most efficient (e.g., often are heavy subsidies).

The PIS region policymakers can learn from these experiences (e.g., what are the most reasonable (i.e., achievable) renewable energy targets for their country, which mix of policy instruments is suitable for their country, what should be the time horizon for fiscal incentives including subsidies, and what non-financial barriers have to be removed for the success of any policy instrument in their country), to enact cost-effective and renewable technology propelling energy policies to meet the CO_2 reduction targets and move towards low-carbon economies.

3.8 Nuclear Power and the Role of Policy Instruments

The development and growth of nuclear power, the cleanest source of electricity generation, for the most part, and good reasons, depending on a country's aspiration and strategic objectives, is in the purview of the public domain. Primarily for the safety and potential weaponization fears, nuclear power projects are built by governments and their related agencies and institutions. However, when these issues are taken care of by the governments, the fundamental questions about the profitability of nuclear power technology still need clarity and answers before any substantial participation and investments by the private sector could be expected.

Policy instruments like RPS have a major role in the promotion of clean energy and nuclear power in a country (NEA, 2004). For instance, enactments of regulations towards the effective and system-wide implementation of the carbon tax and air-quality standards can favor more nuclear power in the electricity supply mix of many countries. A recent example of, fossil-dominant electricity generator, UAE, which has added nuclear power with a notable share of 25% in the total electricity generation of the country, is a welcome move. This new addition of nuclear capacity is substantially reducing the electricity-related emissions in the country. This success of UAE in adding clean electricity generation capacity to its supply mix might encourage Saudi Arabia to bring realization to its stated and aspired clean energy-related initiatives including the building of nuclear power capacity in the country. Inadequate attention to and implementation of these initiatives can increase the long-term costs of energy sector investments and other costs related to the mitigation of CO_2 emissions. On the other hand, both Pakistan and India are adding nuclear power capacity on an ongoing basis (e.g., as of December 2020, the new nuclear power plants under construction or planned for construction in Pakistan and India are 6.1 GW[6] and 2 GW, respectively) (IAEA, 2021).

Subsidies are another important class of policy instruments under a government's regulatory undertakings. The nuclear power sector has been a large beneficiary of government subsidies for its R&D activities (e.g., in the case of OECD countries, over 65% of R&D spending went to nuclear power sector research activities). Nuclear power's contribution as a reliable, low-carbon energy system needs to be appropriately quantified in the regulator and policy frameworks of countries. Such a systematic and policy-neutral treatment of nuclear power can make nuclear power systems more price-competitive with other clean generation technologies. For instance, several advanced countries are investing in advanced technologies and solutions to harness the potential nuclear power to mitigate climate change and reduce their electricity-related CO_2 emissions: (i) France has included nuclear innovation in its national COVID-19 recovery plan with up to €500 million of new investments in key skills development, radioactive waste management, fuel cycle R&D, research facilities and the development of the French small modular reactor (SMR), known as NUWARD™, (ii) the Canadian government has also recently announced funding of over $C70 million to support research and development for SMR technologies in

[6] Each GW of new build is estimated to require US $2–7 billion of investment (IAEA, 2021).

Canada, (iii), The United States has recently launched the Foundational Infrastructure for the Responsible Use of Small Modular Reactor Technology (FIRST) program, The FIRST program, with the initial US $5.3 million investment, which supports capacity-building efforts with partner countries with advanced capabilities, (iv) the UK government with about 17% share of nuclear power in its electricity supply mix, is investing up to £395 million to help develop the next generation of nuclear technologies (IAEA, 2021), and (v) the government of Poland anticipates investment of US $410 billion in the energy transition over 2021–2040 and includes construction of 6–9 GW of nuclear generation capacity to meet increasing electricity demands while achieving climate goals and maintaining stable prices (PMCE, 2021). With these encouraging subsidies and investments in the next-generation nuclear power technologies, it is not unreasonable to hope for a greater share of nuclear power in the overall electricity supply mix of the world.

3.9 Summary of Chapter 3: The Key Insights

Both public and private investments are essential for the development and deployment of renewable and clean energy technologies. As they say, "no one model fits all" when it comes to the enactment, adaption, and implementation of policy instruments, appropriate conservation, and mapping of local conditions and public perceptions regarding renewable energy solutions is critical. The design and implementation of policy instruments are based on the assumption that a strong political will exists in a country. The policy objectives are set first. Then policy instruments are used as a means to achieve those renewable energy policy objectives. There are five major categories of policy instruments: (i) price-based, (ii) quantity-based, (iii) regulations and standards-based (iv) public procurement-based, and (v) auctions-based. The key insights about each of these policy instruments are described here:

i. The RPS is the most popular quota-based policy instrument. The primary aim of an RPS to create a stable and predictable market for renewable electricity that maximizes the benefits of renewable generation while maintaining a lower production cost. The key success factors for RPS-based renewable energy policy are (i) the purchase obligation should be durable and increase with the increased production, (ii) meeting the targets of renewable energy policy should be the focus of the RPS contract, (iii) effective enforcement is needed, (iv) non-compliance penalties should be set higher than compliance cost, (v) consistent monitoring of compliance is required.
ii. When it comes to the design of pricing-based policy instruments, FIT-based renewable energy policies enjoy the most favorable status in the world. FIT policies have various payment models. No one model fits all. The key success factors for FIT-based renewable energy policy include (i) strong project financing support, (ii) cost-effective and efficient procurement procedures, (iii) hedge against project delays, disruptions, and cancellations, (iv) focus on

competitive cost renewables, (v) sound support for emerging technologies, and (vi) provide ratepayer backing.

iii. Regulations and standards-based instruments are mandatory as dictated by the regulatory and legislative body of a country. Regulatory standards can promote renewable energy both directly: by proving incentives for renewable energy and indirectly: by penalizing the competing source of energy such as fossils-based generation. Enforcement, monitoring, and evaluations regularly are critical factors for the success of these standards-based policy instruments of the deployment of renewable energy technologies.

iv. Public procurement-based instruments are leveraged through large procurement and purchasing programs of governments. The government can affect the development of renewable energy markets and technologies by enforcing its suppliers to adhere to creating renewable energy-based production for their supplies. The establishment of public benefits fund can be utilized to spur the development of renewable energy technologies in a country. The clear eligibility of renewable energy technologies and equitable funding support to the developers are important factors in the success of a public befit fund in the promotion of renewable energy.

v. In auctions-based policy instruments, a selection process to allocate goods and services competitively, based on a financial offer. Auctions can be used to determine appropriate tariff rates for a FIT policy and auctions can be a very attractive mechanism for attracting new renewable energy supply in a country by lowering the tariffs.

vi. The development and growth of nuclear power, the cleanest source of electricity generation, for the most part, and good reasons, depending on a country's aspiration and strategic objectives, is in the purview of the public domain.

vii. The nuclear power sector has been a large beneficiary of government subsidies for its R & D activities. With continued governments' support, subsidies, and investments in the next-generation nuclear power technologies, it is not unreasonable to hope for a greater share of nuclear power in the overall electricity supply mix of the world.

The PIS region policymakers can learn from the best practices and adapt to cost-effective renewable and clean energy generating technology propelling energy policies to achieve the much-needed clean electricity. Specially, Pakistan and India have implemented various policy instruments including FITs, RPSs, and subsidies with some successes. They need to reinforce the implementation of these policy instruments to spur the growth of renewable energy in their supply mix. For Saudi Arabia, Although the government is actively promoting several green energy initiatives, concrete policies and regulations (e.g., setting the targets for various clean electricity-generating technologies) are still a work in progress (Elrahmani et al., 2021). Effective energy policies and incentives are no brainer for any country for a successful move towards a low-carbon emitting electricity supply and a low-carbon economy.

References

Azuela, G., & Barroso, A. (2011). *Design and performance of policy instruments to promote the development of renewable energy: Emerging experience in selected developing countries*. The World Bank. https://openknowledgeworldbank.org/handle/10986/9379. Accessed 20 Mar 2019.

Benitez, P. (2012). *Policy instruments for renewable energy: An introduction*. World Bank Institute, Climate Change Practice, World Bank. https://www.esmap.org/sites/esmap.org/files/ESMAP%20IFC%20Re%20Training%20World%20Bank%20Benitez.pdf. Accessed 1 Nov 2019.

Cory, K., Bolgen, N., & Sheingold, B. (2004, March 28–31). *Long-term revenue support to help developers secure project financing* (Paper for Global WINDPOWER 2004).

Cory, K., Couture, T., & Kreycik, C. (2009). *Feed-in tariff policy: Design, implementation, and RPS policy interactions* (Technical Report, NREL/TP-6A2-45549). National Renewable Energy Laboratory.

de Jager, D., & Rathmann, M. (2008). *Policy instrument design to reduce financing costs in renewable energy technology projects*. ECOFYS, Commissioned by the International Energy Agency—Renewable Energy Technology Deployment. http://www.iea-retd.org/files/RETD_PID0810_Main.pdf. Accessed 20 Oct 2018.

del Río, P., & Kiefer, P. (2021). Analyzing patterns and trends in auctions for renewable electricity. *Energy for Sustainable Development, 62*, 195–213.

EA. (2020). Energy Act (EA). U.P.G.A. UK Parliamentary Legislation. 2008. Available online: http://www.legislation.gov.uk/ukpga/2008/32/contents. Accessed 25 May 2021.

EIA. (2009). U.S. Energy Information Administration. Data obtained from the Energy Intelligence Unit (EIU), EIA. www.eia.doe.gov.

Elrahmani, A., Hannun, J., Eljack, F., & Kazi, M. (2021). Status of renewable energy in the GCC region and future opportunities. *Current Opinion in Chemical Engineering, 31*, 100664.

ESMAP. (2010). *Food for thought: Unleashing the potential of renewable energy in India*. SASDE. World Bank.

Georghiou, L., Edler, J., UYarra, E., & Yeow, J. (2014). Policy instruments for public procurement of innovation: Choice, design, and assessment. *Technological Forecasting and Social Change, 86*, 1–12.

GETFIT. (2016). *Global energy transfer feed-in tariff (GET FIT) programme Uganda. IEA/IRENA Renewables policies database*. https://www.iea.org/policies/5636-global-energy-transfer-feed-in-tariff-get-fit-programme-uganda. Accessed 13 Dec 2021.

Hamrin, J., Lieberman, D., & Wingare, M. (2006). *Regulator's handbook on renewable energy programs & tariffs*. Center for Resource Solutions. https://resource-solutions.org/document/regulators-handbook-on-renewable-energy-programs-tariffs/. Accessed 20 Oct 2019.

Haufea, M., & Ehrhart, K. (2018). Auctions for renewable energy support—Suitability, design, and first lessons learned. *Energy Policy, 121*, 217–224. http://www.masstech.org/renewableenergy/green_power/MGPPpaperAWEA.pdf. Accessed 6 July 2017.

IAEA. (2021). International Atomic Energy Agency (IAEA). Nuclear energy for a net zero world. https://www.iaea.org/sites/default/files/21/10/nuclear-energy-for-a-net-zero-world.pdf. Accessed 30 Jan 2022.

India-INDC. (2018). India's Intended Nationally Determined Contributions India-(INDC)—Towards Climate Justice. Government of India Ministry of Environment, Forest and Climate Change. http://moef.gov.in/wp-content/uploads/2018/04/revised-PPT-Press-Conference-INDC-v5.pdf. Accessed 26 Aug 2021.

IRENA and CEM. (2015). *Renewable energy auctions: A guide to design* (pp. 38–41). https://www.irena.org/-/media/Files/IRENA/Agency/Publication/2015/Jun/IRENA_Renewable_Energy_Auctions_A_Guide_to_Design_2015.pdf. Accessed 11 Nov 2019.

Kjöllerström, M. (2008). *Public procurement as a tool for promoting more United Nations' sustainable consumption and production patterns sustainable development innovation briefs*. https://sustainabledevelopment.un.org/content/documents/no5.pdf. Accessed 11 Nov 2019.

Klein, A., Pfluger, B., Held, A., Ragwitz, M., Resch, G., & Faber, T. (2008, October). *Evaluation of different feed-in tariff design options—Best practice paper for the international feed-in cooperation: 2nd edition.* Funded by the Ministry for the Environment, Nature Conservation and Nuclear Safety (BMU). http://www.feed-in-cooperation.org/images/files/best_practice_paper_2nd_edition_final.pdf. Accessed 12 Nov 2018.

Kreiss, J. (2016). *The effect of competition levels on auction outcomes.* http://auresproject.eu/sites/aures.eu/files/media/documents/policy_memo-4_competition_2511016.pdf. Accessed 17 Aug 2017).

KSA-INDC. (2015). *The intended nationally determined contribution of the Kingdom of Saudi Arabia under the UNFCCC.* https://www.ctc-n.org/sites/www.ctc-n.org/files/UNFCCC_docs/ksa-indcs_english.pdf. Accessed 29 Aug 2021.

Landry, E. (2017, June 7). Not all RECs are created equal, Sustainability Roundtable Inc. www.sustainround.com/2017/06/07/not-all-recs-are-created-equal/. Accessed 23 Jan 2022.

Larsen, J., Mohan, S., Marsters, P., & Herdon, W. (2018). *Energy and environmental implications of a carbon tax in The UNITED STATES.* An independent report prepared by Rhodium Group for Colombia SIPA Center on Global Energy Policy. https://energypolicy.columbia.edu/sites/default/files/pictures/CGEP_Energy_Environmental_Impacts_CarbonTax_FINAL.pdf. Accessed 23 Jan 2022.

Lu, Y., Khan, A., Alvarez-Alvarado, S., Zhang, Y., Huang, Z., & Imran, M. (2020). A critical review of sustainable energy policies for the promotion of renewable energy sources. *Sustainability, 12*, 5078. https://doi.org/10.3390/su12125078

Manisalidis, I., Stavropoulou, E., Stavropoulos, A., & Bezirtzoglou, E. (2020). Environmental and health impacts of air pollution: A review. *Frontiers Public Health, 8*, 14. https://doi.org/10.3389/fpubh.2020.00014

Mendonca, M. (2007). *Feed-in tariffs: Accelerating the deployment of renewable energy.* EarthScan. ISBN 9781844074662

Miketa, A., & Saadi, N. (2015). *IRENA. Africa power sector: Planning and prospects for renewable energy, Synthesis Report.*

Ministry of Power India. 2010. http://www.powermin.nic.in/JSP_SERVLETS/internal.jsp. Accessed 11 Jan 2015.

Nadal, S., & Kushler, M. (2000). Public benefit funds: A key strategy for advancing energy efficiency. *The Electricity Journal, 13*(8), 74–84.

NEA. (2004). Nuclear energy Agency (NEA). Government and Nuclear EnergyOECD 2004 NEA No. 5270. https://www.oecd-nea.org/upload/docs/application/pdf/2019-12/5270-government-nuclear-energy.pdf. Accessed 21 Jan 2022.

NTDC. (2018). *Electricity demand forecast based on multiple regression analysis.* National Transmission and Distribution Company, Government of Pakistan. http://www.mocc.gov.pk/SiteImage/Policy/National%20Climate%20Change%20Policy%20of%20Pakistan%20(2).pdf. Accessed 27 Nov 2020.

PAK-INDC. (2016). Pakistan's Intended Nationally Determined Contribution. Ministry of Climate Change, Government of Pakistan. http://mocc.gov.pk/PublicationDetail/ZGM3YjM3MDEtMjcxOS00MDNjLTliMTMtNTk2MmNhZDRiMWMw. Accessed 17 Aug 2021.

PMCE. (2021). Poland Ministry of Climate and Environment (PMCE) Energy Policy of Poland until 2040.

Qudrat-Ullah, H. (2015). Independent power (or pollution) producers? Electricity reforms and IPPs in Pakistan. *Energy, 83*(1), 240–251.

REN. (2021). Renewables 2019 global status Report. https://www.ren21.net/wp-content/uploads/2019/08/Full-Report-2018.pdf, 2018. Accessed 23 Jan 2022.

REN21. (2016). *UN report on renewable energy policy network for the 21st century.* https://www.ctc-n.org/network/network-members/renewable-energy-policy-network-21st-century. Accessed 9 Nov 2019.

Safarzadeh, S., Barzoki, R., & Hejazi, R. (2020). A review of optimal energy policy instruments on industrial energy efficiency programs, rebound effects, and government policies. *Energy Policy, 139*, 111342. https://doi.org/10.1016/j.enpol.2020.111342

Schwerhoff, G., & Sy, M. (2017). Financing renewable energy in Africa—Key challenge of sustainable development goals. *Renewable and Sustainable Energy Reviews, 75*, 393–401.

Uddin, R., Shaikh, A. J., Khan, H. R., Shirazi, M. A., Rashid, A., & Qazi, S. A. (2021). Renewable Energy Perspectives of Pakistan and Turkey: Current Analysis and Policy Recommendations. *Sustainability, 13*, 3349. https://doi.org/10.3390/su13063349

UNIDO. (2009). *Scaling up renewable energy Africa.* 12th Ordinary Session of Heads of State and Governments of the African Union, Addis Ababa.

USAID. (2019). *Developing the net metering market in Pakistan.* SUSTAINABLE ENERGY FOR PAKISTAN (SEP) PROJECT. https://pdf.usaid.gov/pdf_docs/PA00XK9K.pdf. Accessed 27 Jan 2022.

Valasai, D., Uqaili, A., Memon, H., Samoo, R., Mirjat, N., & Harijan, K. (2017). Overcoming electricity crisis in Pakistan: A review of sustainable electricity options. *Renewable and Sustainable Energy Reviews, 72*, 734–745.

WB. (2008). *Development of renewable energy: Emerging experience in selected developing countries* (Energy and Mining Sector Board Discussion Paper No. 22). The World Bank. http://blogs.worldbank.org/climatechange/how-renewable-energy. Accessed 5 Nov 2019.

Wiser, R., & Barbose, G. (2008, April). *Renewable portfolio standards in the United States: A status report with data through 2007*. Lawrence Berkeley National Laboratory technical report LBNL-154E. http://eetd.lbl.gov/ea/EMS/reports/lbnl-154e-revised.pdf. Accessed 20 Mar 2016.

WNA. (2021). World Nuclear Association (WNA). Renewable Energy and Electricity. https://world-nuclear.org/information-library/energy-and-the-environment/renewable-energy-and-electricity.aspx. Accessed 27 Jan 2022.

World Bank. (2018). *Urban Population.* Consulté le October 28, 2019, penetrating The World Bank Data: https://data.worldbank.org/indicator/SP.URB.TOTL.in.zs?view=map. Accessed 30 Oct 2019.

Wyman, O. (2017). *Auctions as a potential mechanism to accelerate green mini-grids in Africa.* A Report for UFID, UK. https://assets.publishing.service.gov.uk/media/59d73990ed915d4937c45b7e/Mini-Grids_Auction_Study_FINAL_FULL.pdf. Accessed 13 Oct 2019.

Zhao, Y., Tang, K., & Wang, L. (2013). Do renewable electricity policies promote renewable electricity generation? Evidence from panel data. *Energy Policy, 62*, 887–897.

Ziras, C., Calearo, L., & Marinelli, M. (2021). The effect of net metering methods on prosumer energy settlements. *Sustainable Energy, Grids and Networks, 27*, 100519.

Chapter 4
Simulation and Modeling in Service of Energy Systems

4.1 Introduction

Energy systems are complex, dynamic systems affecting human life economically (i.e., production and consumption of products and services), socially (e.g., communication, welfare, and well-being), and environmentally (i.e., emissions, healthcare, and sustainability). Generally, understanding of a system, such as an electricity system, precedes any actions or decisions (e.g., energy policy design and implementation). Given the rising global concerns about climate change and power sectors' environmental emissions (Manisalidis et al., 2020; Uddin et al., 2021), a better understanding of the dynamics of the energy system: how do the affordability, reliability, and cleaner production of electricity for a nation or region interact over time to unfold the socio-economic and environmental realities of our world, is an ongoing research endeavor. Also, due to the rapidly decreasing, cost of solar and wind power-based electricity generation technologies, global investments in renewable energy solutions are on the rise (IRENA, 2020), countries of the PIS[1] region are heavily dependent on fossil-based electricity generation, which can also benefit from such clean energy solutions. Policymakers in the PIS region that are, therefore, face a challenge in having a balanced energy supply mix in their countries.

The modeling and simulation community has long been supporting the policymakers in their understanding and decision making in and about various complex systems including energy systems. In particular, system dynamics (Forrester, 1961) models have been developed and applied successfully to address a variety of issues (e.g., privatization and deregulations of electricity systems, analysis, and design of various theme-based electricity supply mix (e.g., meeting CO_2 emissions reduction targets, the multi-dimensional impact analysis of policies of the regions and countries, and design and evaluation of various policy incentives and mechanism) in the energy domain. In this chapter, we present:

---

[1] PIS: Pakistan, India, Saudi Arabia.

H. Qudrat-Ullah, *Understanding the Dynamics of Nuclear Power and the Reduction of CO_2 Emissions*, https://doi.org/10.1007/978-3-031-04341-3_4

i. an overview of electricity supply and CO_2 emission related modeling and simulation studies in the context of the PIS region,
ii. to what extent system dynamics-based modeling has contributed to the understanding and decision making related to the power sector related CO_2 emission, and
iii. have identified the research gap that the research in this book is aimed to fill.

We propose System Dynamics (SD) modeling approach that can systematically account for the dynamics of energy, economy, and environment holistically. By utilizing this modeling framework, we expect to build and test multiple renewable and nuclear energy focussed scenarios—the key input to policymaking for having cleaner energy production and the reduction of CO_2.

4.2 Electricity Supply and CO_2 Emissions Modeling in the PIS Region

Global concerns, initiatives, and policies of various regions and countries about CO_2 emissions provide the impetus for renewed research on electricity-related CO_2 emissions and their impacts. Given the nature of complexities (e.g., electricity supply systems are dependent upon decisions by various stakeholders, and the constant need for reliable and effective management of diverse technologies for electricity production and large distribution networks) and uncertainties (e.g., demand of electricity, fuel price, and weather-related uncertainties are abound in the electricity supply systems), it is not surprising to see a diverse set of research conclusions in this regard. For instance, in the case of India, some argue (e.g., Ru du Can et al., 2019) that the current course of actions and policies will enable India to meet its INDC's[2] commitment regarding CO_2 emissions. Others (e.g., Shearer et al., 2017; Tiewsoh et al., 2019) predict higher demand for electricity and much higher CO_2 emissions. Table 4.1 lists the major conclusions of the most recent studies related to electricity supply and CO_2 emissions of the PIS region.

Overall, when it comes to modeling and evaluation of electricity-related CO_2 emissions in the PIS region, various modeling approaches, predominantly the econometric and linear programming-based models have been applied. The main takeaway from these empirical studies about power-related CO_2 emissions in each of the PIS region countries are: (i) Pakistan should focus on increasing renewable and cleaner production of electricity into its supply mix, (ii) India needs stringent measures to control rising electricity-related CO_2 emission, and (iii) Saudi Arabia should undertake clean-energy use in the transportation sector to off-set rising CO_2 emissions due to larger oil-based consumption and economic activities in the country. It would be interesting to systematically assess the current energy policies and initiatives of these countries to see whether they can achieve both short-term (e.g., meeting the targets

[2] INDCs: Intended Nationally Determined Contributions.

Table 4.1 Key recent studies about electricity and CO_2 emissions modeling in the PIS region

Country	Modeling approach	Key findings	Reference studies
India	LEAP[a]	India can achieve its INDC[b] target with currently implemented policies	Rue du Can et al. (2019)
Pakistan	Simulation	With increasing carbon taxes, the LCOE[c] decreases, and consequently, the diesel fuel consumption increases	Jahangir et al. (2021)
KSA	Nonlinear co-integration	Non-oil income and urbanization lead to an increase in CO_2 emissions per capita	Mahmood et al. (2020)
Pakistan	Review study	Biomass energy can help lower CO_2 emissions and increase the share of renewables by 5% in 2030	Irfan et al. (2020)
Pakistan	LMDI[d]	Decoupling between Pakistan's transport CO_2 emissions and GDP is needed	Raza and Lin (2020)
Pakistan	ARDL[e]	Nuclear power has an insignificant effect on CO_2 emissions	Usman et al. (2020)
Pakistan	TIMES[f]	Coal-based electricity generation would be a major source of emissions leading to the highest amount of air pollution	Rehman et al. (2020)
KSA	Simulation	On average, for each 1% of EV deployed, emissions would reduce by 0.5% and in the best-case emissions reduce by 0.9%	Amro and Peerbocus (2020)
Pakistan	Systematic review	CO_2 Emissions mitigation can be achieved through CCS[g] technologies	Rashid et al. (2020)
Pakistan	Least square estimator	The inverted U-shaped Environmental Kuznets Curve hypothesis for per capita income and per capita CO_2 emissions in Pakistan are not supported	Naz et al. (2019)
India	LEAP	A two to a three-fold increase of the electricity demand in all scenarios by 2030	Tiewsoh et al. (2019)

(continued)

Table 4.1 (continued)

Country	Modeling approach	Key findings	Reference studies
Pakistan	LEAP	The government's power expansion plans will emit relatively more CO_2 emissions	Mengal et al., (2019a, 2019b)
India	Johansen co-integration	India should take stringent measures to curb the surging CO_2 emissions	Pandey and Rastogi (2019)
Pakistan	NARDL[h]	More nuclear or renewable energy technologies can help in the eradication of CO_2 emissions	Luqman et al. (2019)
India	Interpretive analysis	India's per capita emissions in 2030 will remain well below today's global average	Dubash et al. (2018)
KSA	ARDL	Energy use increases pollution both in the short and in long run in the country	Raggad (2018)
India	Survey approach	Emissions from the proposed coal plants would exceed the country's climate commitment to reduce CO_2 emissions	Shearer et al. (2017)
KSA	ARDL	Higher technological progress in the production process would reduce CO_2 emissions without harming economic growth	Samargandi (2017)

[a] LEAP: Long-range Energy Alternative Planning System
[b]INDC: Intended National Determined Contribution
[c]LCOE: The Levelized cost of energy
[d]LMDI: Logarithmic Mean Divisia index
[e]ARDL: Autoregressive Distributed Lag
[f]TIMES: The Integrated Markal-Efom System
[g]CCS: Carbon Capture and Storage
[h]NARDL: Non-linear Autoregressive Distributed Lag
From 2017 to 2021; *Source* adapted from Qudrat-Ullah (2022)

sets in their INDCs by 2030) and long-term goals (e.g., achieving low-carbon and sustainable economies of their countries).

4.3 System Dynamics Modeling for Energy Systems

To better understand the dynamics of the electricity supply system and the complex task of reduction of environmental emissions, policymakers need and often use a variety of decision support systems and tools. For the most part, the traditional modeling methods and approaches (e.g., ARLD, LMDI, TIMES, MARKAL, LEAP, input-output, linear programming, optimization, surveys, econometric, simulation, and review-based analyses) have served the energy domain well (Akbar et al., 2017; Dyner & Bunn, 1997; Mengal et al., 2019a, 2019b; Nasirov et al., 2020). However, for addressing the highly dynamic and complex problem, "understanding the dynamics of cleaner electricity production and reduction of CO_2 emissions," requires accounting for (i) the system-wide structures of the electricity systems (e.g., feedback loops, delays, non-linearities, and multiple objectives, often conflictive, of the stakeholders) responsible for the dynamic behavior of the system (e.g., the dynamics of electricity supply mix and CO_2 emissions) (ii) the PA's[3] targets, simultaneously, where these traditional approaches can hardly succeed. They lack on several fronts including (Qudrat-Ullah, 2015a, 2015b, 2022):

- The non-linear relationships among various variables of the electricity system (e.g., a slight increase in the electricity price may not lead to any decreased use of electricity by the users, however, if electricity prices continue to rise then some users may use relatively less of electricity) are rarely modeled.
- The dynamic interactions (and feedback) among the variables of an electricity system are usually approximated with uni-directional relationships.
- They barely distinguish between and account for the physical delays (e.g., in the transportation of power plants' machinery to the site) and the information lags (e.g., in the approval of a power plants' site) explicitly in their models.
- The elasticity of substitution between the competing electricity generation technologies is often assumed constant.
- Perceptions of people (e.g. regarding the adoption of renewable technologies and modern devices) are not explicitly modeled.
- The desired and actual state of variables of the electricity system (e.g., desired and actual stock of production capital) are rarely distinguished in their models.
- Non-linear responses to actions (e.g., the effect of rising electricity prices on industrial and commercial consumption) are not explicitly recognized and modeled.
- They inadequately represent the interactions between energy, economy, regulatory frameworks, and the environment—the core structural elements of an electricity system and an essential requirement for any economy-wide analysis of an energy policy.

To overcome these shortcomings of various modeling approaches, the system dynamics approach (Forrester, 1962; Sterman, 2000), stands out for its strengths and

[3] PA: Paris Agreement.

abilities including to (Al-Sarihi & Bello, 2020; Qudrat-Ullah, 2015a, 2015b, 2017, 2022):

i. capture the essential feedback structures underlying any energy system,
ii. account for the endogenous complexity of energy systems (e.g. internal to model electricity demand generation, prices, emissions) which is considered responsible for the behavior of the system,
iii. ability to model non-linearities and soft variables (e.g., job satisfaction level of the plant operators in the control room of a power plant or the attitude of users toward the adoption of renewable technologies),
iv. link the observable patterns of a system (e.g., power-related CO_2 emissions) to micro-level structures and decision-making processes (e.g., investments in renewable energy technologies), and
v. explicit and distinctive modeling of perceptions (e.g., users' perception about the utility of clean-energy appliances and technologies), 'desired' and the 'actual' states of the various variables (e.g., additional capacity of clean-energy-generating technologies) of an electricity system.

System dynamics (SD) models have been developed and applied to a range of energy-related issues including national energy policy design and evaluation, energy investments and uncertainty, inter-fuel substitution in OECD[4]—European electricity production, privatization of the electricity industry, power-related and emissions. Table 4.2 provides a sample of SD models that have been used to address a variety of electricity-related issues across the countries, over time.

For instance, Ford (1997), in an award-winning article in the 1996 International System Dynamics Conference, documents and summarizes the impressive body of work that system dynamics practitioners have accumulated over the past several decades. In particular, he gives his interpretation of the important and unique features of the system dynamics approach and its contribution to the electric power sector (Ford, 1997). This seminal work by Ford provides a theoretical foundation for the development of our dynamic model, which is illustrated in Chapter 4 of this book. A similar research attempt, to support energy policy modeling and decision making (Dyner & Bunn, 1997), introduces an integrated system dynamics-based framework. Our modeling framework, which is described in this chapter, Chapter 4 of this book, benefits from the various modeling structures (e.g., liberalization sector feedback loops, and delays) of this article.

Park et al. (2007), using SD modeling, analyzed the effects of various capacity incentive systems on new investment in the Korean electricity market. In an alternative, explicit formulaic approach, such as linkage to the periodic loss of load probability (LOLP), a direct comparison with fixed payments is carried out. In their simulation-based results, they suggest that instead of staying the course with a fixed capacity payment approach, a LOŁP-based capacity mechanism may, in the longer term, help to achieve resource adequacy and increase the reserve margin (Park

[4] OECD: Organization for Economic Co-operation and Development.

Table 4.2 A sample of SD models in service of energy-related issues

Reference studies	Key focus/Issues	Country/Region
Qudrat-Ullah (2013)	Electricity generation capacity	Canada
Laimon et al. (2020), Park et al. (2007)	Energy investments and uncertainty	Korea; Australia
Moxnes (1990), Abada et al. (2013), Gomez et al. (2017)	Inter-fuel substation	OECD-Europe; OECD; Chili
Dyner and Bunn (1997), Gholizad et al. (2017), Ford (1997)	Privatization of the electricity industry	UK; Iran; US; Colombia
Saysel and Hekimoglu (2013), Laimon et al. (2020), Qudrat-Ullah (2022)	Power-related emissions	Turkey; Australia; Pakistan
Sisodia et al. (2016), Panda (2011), Qudrat-Ullah (2022)	Energy demand fulfillment; Impact of renewable energy sources	India; Pakistan
Naill (1992), Laimon et al. (2020), Qudrat-Ullah (2015a, 2015b, 2022)	National energy policy design and evaluation	US; Australia; Pakistan
Shafiei et al. (2015)	Alternative fuel markets	Iceland

et al., 2007). The reserve margin component of our dynamic model utilizes a similar formulation as is described in this paper.

Renewable energy plays a critical role in the energy mix of a country that is aiming to reduce and mitigate power sector emissions. Utilizing a system dynamics simulation model, Panda (2011) has evaluated the renewable resources in India as a potential non-filled generation source of electricity. His dynamic model-based analysis indicates that:

i. Even with a conservatively estimated per capita electricity need of 2000 kWh/annum, India would need to generate approximately 3000 TWh/yr. However, analysis of available information about available renewable sources, India can only produce around 1000 TWh/yr, and
ii. Renewable sources alone will not suffice for meeting India's future electricity needs (Panda, 2011).

Characterizing the electricity demand and supply system of Canada as a complex and dynamic system, Qudrat-Ullah (2013) provides a detailed description of the system dynamic model that is used to analyze the dynamics of the generation capacity of the electricity system of Canada (Qudrat-Ullah, 2013). His model-based scenario analysis finds that in addition to the traditional generation capacity and supply adjustment methods, substantial new investments in electricity generation capacity and efficiency enhancement areas are needed to achieve a sustainable and balanced electricity supply and demand system in Canada.

For most of the countries of the world, the electricity supply sector of their economies is often the largest CO_2 emitting sector. On the other hand, the electricity sector presents a great opportunity for nations to combat or mitigate CO_2 emissions. In the case of Turkey, with increasing demand for electricity, power sector CO_2 emissions had tripled over the last two decades (Saysel & Hekimoglu, 2013). In their study, Saysel and Hekimoglu (2013), they describe the building and validation of a system dynamic model to analyze the alternatives for the mitigation of electricity-related CO_2 emission in Turkey, Saysel & Hekimoglu, 2013). Their SD model comprises structures that represent (i) investment, (ii), dispatch and pricing heuristics, and (iii) the natural resource base of electricity supply in Turkey. Their model-based scenario analysis indicates that there are policy options including feed-in-tariffs, investment subsidies, and carbon taxes that can reduce electricity-related CO_2 emission below 50% of the business as usual growth. A similar model-based scenario analysis approach is expected to be developed and applied for exploring the electricity-related CO_2 emissions pathways for the case of Pakistan.

One of the distinguishing features of SD modeling is to account for the dynamic nature of the inter-fuel substitution mechanism in an energy supply system. Moxnes (1990) was the first researcher who, by building an innovative SD mode, demonstrated the various dynamics of inter-fuel elasticity of substitution process (Moxnes, 1990). Overcoming the traditional simplistic representation of the energy demand function by a single-linear, linear, and inverse demand function that does not account for any dynamics of energy substitution, Abada et al. (2013) introduce an enhanced functional specification of inter-fuel substitution function using the system dynamics-based model. Using their dynamic model, they (i) confirm the pertinence of this modeling framework to represent inter-fuel substitution at different fuel prices in the industrial sector, and (ii) propose a dynamic functional specification of the demand function for natural gas (Abada et al., 2013). A similar inter-fuel substitution process for electricity-generating technologies is to be constructed in our dynamic model.

Energy systems are complex systems, at best. This complexity of the energy systems primarily comes from the existence and interactions of non-linear and dynamic variables including various stocks of electricity generation capacity, restricting and regulatory regimes, fuel supply, and price dynamics, and advances and challenges in technologies for electricity generation, transmission, and consumption. System dynamics modeling is capable of accounting for and dealing with the complexity of energy systems (Sterman, 2000). To better understand the dynamics of such a complex system—the demand and supply dynamics of the electricity industry of Canada—Qudrat-Ullah (2013) develops, validates, and utilizes a system dynamics-based simulation model. Based on the dynamic model-based analysis, he concludes:

i. Instead of continuing with the traditional electricity demand–supply adjustment methods, Canada should adopt new energy strategies primarily focused on capital assets. productivity, and efficiency to avoid a downward spiral of electricity industry deficiency, and

ii. With an additional investment of about 10 billion CAN$ over a decade (2015–2025), Canada can effectively manage its electricity demand–supply dynamics with relatively cleaner production in its electricity supply mix, and the Canadian economy can expect better energy intensity (0.21versus 0.25 toe/million$), resulting in global and convincible recognition of Canada as a green economy.

The transition process towards alternative fuel markets is a complex, dynamic task. Utilizing a system dynamics model, Shafiei et al. (2015) explore the transition process towards alternative fuel markets in the case of Iceland. Their dynamic model takes into account the entire energy system including interactions and feedback among various sectors including (i) supply sectors, (ii) energy prices, (iii) infrastructure, and (iv) fuel demand. They apply their model to simulate the transition path towards the alternative fuel market during the time horizon of 2015–2050 in the context of Iceland. They analyze the transition pathways towards hydrogen and biofuel markets in Iceland and discuss implications for the alternate fuel demand, fuel prices, and needed fuel-supply infrastructure. They conclude that:

i. the dynamic model simulation effectively exhibits the continual transition towards equilibrium as market prices dynamically adjust to changes in supply and demand,
ii. the developed model can also simulate the impact of different policy instruments on both supply and demand sides,
iii. the developed dynamic model can illustrate how the decisions made by energy suppliers and infrastructure owners influence consumer behavior,
iv. the developed dynamic model can also be used to understand the impact of demand for alternative fuels on the development of energy infrastructure can be evaluated, and
v. if the government regulates the wholesale prices for alternative fuels or provides energy supply infrastructure, the fuel supply decisions should be treated differently (Shafiei et al., 2015).

According to UN's SDG[5] No. 7 "ensure affordable, reliable, and cleaner electricity for all." Decision makers and energy policymakers of developing countries are striving hard to meet this goal. Pakistan, a developing nation with a rising population and electricity demand, started the privatization of its energy sector in the early 1990s. Qudrat-Ullah (2015a, 2015b) examines Pakistan's experience with electricity reforms about the various energy policies enacted in Pakistan over the past two decades. Utilizing a dynamic model-based scenario analysis approach, he analyzes the effects of privatization in general and IPPs' investments in Pakistan's electricity sector in particular. His major findings are:

i. contrary to the generally perceived positive effects of privatization, liberalization, and IPPs' energy sector investment, Pakistan's electricity sector is faced with severe gaps between demand and supply, rising electricity rates, and increasing electricity-related CO_2, SO_2, and NO (nitrogen oxide) emissions,

[5] SDGs: Sustainable Development Goals.

ii. in the case of Pakistan's experience with IPPs, due to their major investments in oil-fired power plants—quick earning technology choice, are the major electricity-related pollution producers,
iii. an indigenous resource-focused alternate energy policy option appears to provide cheaper and cleaner electricity in Pakistan,
iv. the existing Energy Policy of Pakistan falls short on at least three fronts: (i) it is not environmentally friendly, (ii) consumers have to face higher rates for a long period, (iii) the load-shedding and unpredictable shortages of electricity does not appear to close in a balanced manner shortly,
v. with the "demand–supply imbalances" at hand, Pakistan has the opportunity to adjust and re-design the existing Energy Policy,
vi. substitution of indigenously resource-rich and relatively low-carbon solutions, such as in our alternate scenario, could result in relatively cheaper and cleaner electricity in Pakistan,
vii. a plan like our alternate plan can also make Pakistan's supply mix import-independent,
viii. the utilization of hydro reservoirs as "energy storage systems" appears to be a cheaper source of "flexible supply" than a thermal generation. With an increasing share of hydropower plants, CO_2 emitting thermal generation can be eliminated from the supply mix. Thus, resulting in more affordable "cleaner electricity" for Pakistanis (Qudrat-Ullah, 2015a, 2015b).

Energy demand fulfillment is the prime task for energy policymakers across the globe. Although fossil-based electricity generation has served the world well and for a long period, staying the course is not an option for energy policymakers. In the case of India, perhaps the first-ever study to use qualitative system dynamics methodology in the domain of energy demand fulfillment, Sisodia (2016) explore the role of solar power as an alternate source of energy for India (Sisodia, 2016). He proposes the involvement of stakeholders as a prerequisite for the adoption and successful implementation of any renewable energy policy in India. This conclusion is consistent with other studies where active "engagement' of the community and stakeholders is seen as a critical factor for the success and ownership of renewable energy-based solutions, devices, and appliances (Qudrat-Ullah et al., 2020).

Successful privatization of the power sector requires appropriate conditions and policy measures including successful implementation of market competition and the elimination of exclusive market conditions. Without active and direct government support, the private sector is not able to reasonably manage this area of economy and industry (Gholizad et al., 2017). Therefore, the theme of privatization and liberalization of the power sector of the economy, due to the inherent complexity of this problem, has received due attention from the system dynamics modeling community. In a case study of Iran's power sector, by developing, validating, utilizing a system dynamics model, Gholizad et al. (2017) assess the impacts of deregulation on generation capacity growth as well as expected annual operating costs and revenues. They consider the development of a novel system dynamic model to analyze the effects of capacity investment on the performance measures in the electricity market as the core contribution of this study. Their model of the electricity market includes supply (capacity), demand and energy market sectors and enables the assessment of the effect of prices on average consumption. They draw several conclusions and policy insights including:

i. the process of restructuring the electricity industry is of the serious need of the industry to increase productivity and attract appropriate private investment partnerships,
ii. for effective policy implementation, It is necessary to distinguish between electrical and electrical goods,
iii. if current trends continue, the investment need for the industry cannot be supplied from domestic sources alone, and
iv. the government must support the private sector in the field of infrastructure investment and provide tools and incentives for the private sector (Gholizad et al., 2017).

The role of governments in supporting and developing the private sector for the power sector is critical. In addition to the costly development of the needed infrastructures (e.g., distribution networks), support of governments and related departments across various levels (municipal, provincial, and national) in the promulgation of energy policy instruments (e.g., feed-in-tariffs, RPS, power purchase agreements, tax incentives, etc.) to attract investments and for the sustainable operations of IPPs and other private generators is needed on continues basis. In the case of developing countries where electricity markets are new or just being built, the need for governments becomes even more pronounced.

Building on the works of Moxnes (1990) and Abada et al. (2013) to explore a country's energy policy options, Gomez et al. (2017) describe the construction, calibration, and application of a system dynamics-based dynamic energy-planning model that represents the energy matrix behavior (Gomez et al., 2017). Their dynamic model shows the dynamic substitution between energy sources, considering the differing effects of strategic decisions that affect the structure of the energy supply mix. They analyze the evolution of the Chilean energy supply mix, like Chile, a heavily fossil-import dependent country, needs to diversify its energy sources because it is heavily dependent on foreign energy supplies. Their developed model is quite generic in its nature and has been calibrated to the historical demand prices data of the Chilean energy sector. They specifically investigate the research question: whether the current energy policy reforms and incentives in Chile will help reduce the dependence on imported fuel and reduce environmental impacts in the medium to long term (Gomez et al., 2017). Their model-based scenario analysis provides several insights about the introduction of renewables in the energy supply mix of Chile. These insights and conclusions include:

i. compared with the reference year, 2012, the "business as usual" scenario shows a trend where the current heavy dependence on imported fossils continues and there is no substantial change in the energy supply mix,
ii. current energy policy incentives show no change in the energy supply mix towards the stated goals of the government: (i) greater energy supply security,

(ii) reduced dependence on imported fossils, and (iii) less carbon-emitting production,

iii. changes in fuel prices and investment costs (in our simulations) do not produce significant changes from the reference scenario (i.e., with no changes).

iv. the market alone may not be able to bring about the desired change in the composition of the energy matrix in Chile but active intervention by the government is needed (Gomez et al., 2017).

Overall, these results and insights are consistent with our above-mentioned analysis of the role of government in the development and promotion of renewables (e.g., Gholizad et al., 2017)—*privatization and liberalization of the power sector alone are not enough, active governments' interventions, regulations, and policy incentives are needed.*

The key strength of system dynamics methodology, as compared with traditional econometric methods, is its ability to account for and link micro-level decision variables (e.g., investments and regulatory decisions) with the macro-level outcomes of complex systems such as energy systems (e.g., evolution energy supply mic and CO_2 emissions trajectory). Recognizing and building on the strengths of system dynamics, Laimon et al. (2020) develop and apply a system dynamic-based dynamic model to examine CO_2 emissions trends through different possible scenarios. They calibrated their dynamic model with Australian case data. They specifically assessed the current policy scenario with two alternate scenarios. They examined the impact of these policy scenarios on fossil fuel reserves, energy dependency, energy prices, energy bankruptcy, and CO_2 emissions. Their key insights for the energy policymakers are:

i. although for a sustainable energy future, establishing the balance of supply–demand, conservation of resources, and reducing energy dependency and emissions is crucial, the current trend of the Australian energy sector is in line with an unsustainable future and the growth is not being controlled,

ii. the limits to growth are approaching due to excessive fossils extraction, high environmental emissions, and high energy dependency—which could make the current scenario one of the worst scenarios for the Australian energy sector,

iii. in the case of Australia, reducing dependency on fossil fuel and accelerating the transition to full renewable systems could be the best scenario

iv. various policy initiatives and measures including improving energy efficiency, switching to renewable transportation, switching to renewable electricity, electrification of sectors that do not run on electricity by renewable energy could achieve a low-carbon producing energy supply mix for Australian energy sector.. However, more research is required to examine the potential impact of such improvements on the energy sector, which is the topic of the next paper (Laimon et al., 2020).

As we saw in the aforementioned studies, in the case of developing countries, the role of governments in the promotion of renewables and cleaner sources of electricity production is critical, the same appears to be the case with a developed country like Australia. Therefore, no one model fits all. Both public and private sector partnerships are needed to spur sustained deployment of renewable and cleaner energy-producing solutions and technologies and mitigate climate change and global warming effects across the world.

Continuing with the theme of the application of a model-based scenario approach for energy policy assessment and re-design, Qudrat-Ullah (2022) describes the development, validation, and application of a dynamic model to evaluate the evolving electricity supply mix under the current energy policies and support mechanisms and suggest a cleaner and affordable alternative. He constructed and evaluated three energy policy scenarios: (i) current policy incentives and the government's existing regulations and planes are captured in the base case scenario, (ii) the two alternate scenarios focus on the deployment of domestically available renewable resources (i.e., hydro, solar, wind), planned CO_2 taxes, and the government of Pakistan's plan for the expansion of nuclear power. Based on the modeling-based analysis, the key findings and conclusions are:

i. the government of Pakistan's exiting energy policy and plans appears to miss the target of CO_2 emission reduction by 2030,
ii. the alternate scenario focused on the use of the domestic renewable source for electricity generation can help Pakistan to meet its power-related CO_2 emission target, and
iii. compared with the reference scenario, the consumers of electricity in Pakistan appear to pay substantially less (i.e., a saving of about 23% per annum from the year 2020 to the year 2035) under the alternate scenario which is focused on cleaner production of electricity.

These results are consistent with the findings of a recent study that uses the traditional econometric method (Shahzada et al., 2021). It is interesting to note that the results of this study motivate us to embark on this book project. As we will recognize and show in Chapter 4 and Chapter 5, this book heavily draws on this article.

4.4 Identification of the Research Gap

In these SD modeling-based studies, given in Table 4.2, various electricity-related issues including privatization, inter-fuel substitution, demand, energy sector investments, generation capacity, and related environmental emissions, across the regions and countries, are well represented. Privatization and deregulation studies of the early 1990s, the focus of modeling has shifted towards the assessment and design of energy policies for sustainability and dealing with power-sector emissions. It is interesting to note that the energy sectors of both Pakistan and India have some SD modeling-based analysis and policy assessment work, no SD-based studies are found for the case of Saudi Arabia. Therefore, it seems imperative for the policymakers of the KSA to avail of SD-based evaluation and analysis of their energy policy and initiatives within the context of socio-economic and environmental perspectives.

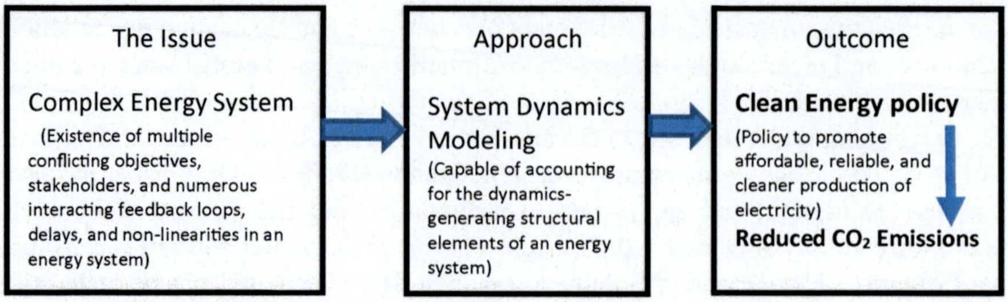

Fig. 4.1 Modeling approach for cleaner energy policy design

In the case of India, prior studies have often reported conflicting results regarding power-related CO_2 emissions. For instance, in an SD-based study, Panda has concluded that renewable sources of energy have will play a significant role in India's supply mix to achieve sustainability (Panda, 2011). On the other hand, in Panday and Rastogi's study, the results indicated (i) for a short run causality from electricity consumption to economic growth and the CO_2 emissions, and (ii) that India should take stringent measures to curb the surging CO_2 emissions (Panday & Rastogi, 2019).

In the case of Pakistan, prior studies, using various econometric modeling approaches, have investigated several energy-CO_2 emissions-related issues including the role of renewable including biomass energy, CCS[6] technologies, and nuclear power in the mitigation and reduction of power sector related CO_2 emissions. Only recently, Qudrat-Ullah (2022) has evaluated the socio-economic and environmental impact of the energy policy of Pakistan using an integrated dynamic modeling approach, system dynamics (Qudrat-Ullah, 2022). Although the role of non-thermal power in the reduction of CO_2 emissions was explored, the long-term impact of increased nuclear generation in CO_2 emissions reduction and achieving a low carbon economy of Pakistan still needs to be investigated. What could be the optimal nuclear power and renewable energy-based energy supply mix of Pakistan? Given the ongoing global concerns, efforts, and regulatory regimes regarding CO_2 emissions, a detailed system-wide assessment, and design of policy instruments to meet the desired CO_2 emissions reduction targets is needed. Therefore, utilizing the research framework shown in Fig. 4.1, in the following chapters, we will describe the development, validation, and application of a system model (SD) for the policy assessment and design in the case of Pakistan. We hope that the molding community and policymakers of India and Saudi Arabia, by utilizing this sound modeling framework can systematically assess and re-design their energy policies.

[6] CCS: Carbon Capture and Storage.

4.5 Summary of Chapter 4: The Key Insights

When it comes to energy policymaking to provide affordable, reliable, and cleaner energy supply to the people of a country, understanding the multidimensional dynamics of energy supply (e.g., nature of supply mix, incentives, and policies for cleaner production and use of electricity, current and future state of power-related environmental emissions) is an essential pre-requisite. The key insights are:

- To aid stakeholders and policymakers in the energy domain, various methods and approaches have been used for several decades.
- For a long-term assessment and design of an energy policy, the development and use of simulation models in general and system dynamics, in particular, is ever on the rise.
- Various methods, techniques, and approaches (e.g., econometric, and system dynamics) in service of understanding the dynamics of various energy-related issues including the electricity-related CO_2 emissions in the PIS region are used.
- The specific and unique strength of the SD approach to link the macro-level issues of energy systems (e.g., reduction of CO_2 emissions) with its micro-level structures (e.g., investments in clean electricity-generating technologies) makes it stand out as the most suitable modeling approach for complex energy systems.
- The energy sectors of both Pakistan and India have some SD modeling-based analysis and policy assessment work, no SD-based studies are found for the case of Saudi Arabia.
- In the case of India, prior studies have often reported conflicting results regarding power-related CO_2 emissions.
- As the focus of this book is on Pakistan's case, the need for a systematic SD-based assessment and design of the energy policy of Pakistan is established.
- The use of system dynamics modeling framework is presented as a viable solution for policymaking for achieving the goal of affordable, reliable, and cleaner production of electricity for any country or region.

References

Abada, I., Briat, V., & Massol, O. (2013). Construction of a fuel demand function portraying inter-fuel substitution, a system dynamics approach. *Energy, 49*, 240–251.

Akbar, M., Thaheem, M., & Arshad, H. (2017). Life cycle sustainability assessment of electricity generation in Pakistan: Policy regime for a sustainable energy mix. *Energy Policy, 11*(1), 111–126.

Al-Sarihi, A., & Bello, A. (2020). Socio-economic and environmental implications of renewable energy integrity in Oman: Scenario modeling using system dynamics approach. In H. Qudrat-Ullah & A. Kayal (Eds), *Climate change and energy dynamics in the Middle East*. Springer.

Amro, M., & Peerbocus, N. (2020). Electric vehicle deployment and carbon emissions in Saudi Arabia: A power system perspective. *The Electricity Journal, 33*(6), 106774.

Dubash, K., Khosla, R., Rao, D., & Bhardwaj, A. (2018). India's energy and emissions future: An interpretive analysis of model scenarios. *Environmental Research Letters, 13*(7), 074018.

Dyner, I. (2000). Energy modelling platforms for policy and strategy support. *Journal of the Operational Research Society, 51*(2): 136–144.

Dyner, I., & Bunn, D. (1997). A simulation platform to analyze market liberalization and integrated energy conservation policies in Colombia. *Energy Policy,* 259–271.

Ford, A. (1997). System dynamics and the electric power industry. *System Dynamics Review, 13,* 57–85.

Forrester, J. W. (1961). *Industrial dynamics.* MIT Press.

Forrester, J. W. (1962). *Industrial dynamics.* MIT Press.

Gholizad, A., Ahmadi, L., Hassannayebi, E., Memarpour, M., & Shakibayifar, M. (2017). A system dynamics model for the analysis of the deregulation in electricity market. *International Journal of System Dynamics Applications, 6*(2), 1–30.

Gomez, C., Arango-Aramburo, S., & Larsen, E. (2017). Construction of a Chilean energy matrix portraying energy source substitution: A system dynamics approach. *Journal of Cleaner Production, 162,* 903–913.

IRENA. (2020). *Global landscape of renewable energy finance, 2020, international renewable energy agency.* Abu Dhabi. https://www.irena.org/-/media/Files/IRENA/Agency/Publication/2020/Nov/IRENA_CPI_Global_finance_2020.pdf. Accessed 30 Oct 2021.

Irfan, M., et al. (2020). Assessing the energy dynamics of Pakistan: Prospects of biomass energy. *Energy Reports, 6,* 80–93.

Jahangir, M., Mostafaeipour, A., Habib, H., Saghaei, H., & Waqar, A. (2021). Effect of emission penalty and annual interest rate on cogeneration of electricity, heat, and hydrogen in Karachi: 3E Assessment and sensitivity analysis. *Journal of Engineering*, Article ID 6679358, 1–16. https://doi.org/10.1155/2021/6679358. Accessed 30 July 2021.

Laimon, M., Mai, T., Goh, S., & Yousuf, T. (2020). Energy sector development: System dynamics analysis. *Applied Sciences, 10,* 134.

Luqman, M., Ahmad, N., & Bakhsh, K. (2019). Nuclear energy, renewable energy and economic growth in Pakistan: Evidence from a non-linear autoregressive distributed lag model. *Renewable Energy, 139,* 299–1309.

Mahmood, H., Alkhateeb, Y., & Furqan, M. (2020). Oil sector and CO_2 emissions in Saudi Arabia: Asymmetry analysis. *Palgrave Communications, 6,* 88.

Manisalidis, I., Stavropoulou, E., Stavropoulos, A., & Bezirtzoglou, E. (2020). Environmental and health impacts of air pollution: A review. *Frontiers in Public Health, 8,* 14. https://doi.org/10.3389/fpubh.2020.00014

Mengal, A., Mirjat, H., Walasai, D., Khatri, A., Harijan, K., & Kaili, A. (2019a). Modeling of future electricity generation and emissions assessment for Pakistan. *Processes, 7*(212), 1–26.

Mengal, A., Mirjat, H., Walasai, D., Khatri, A., Harijan, K., & Kaili, M. (2019b). A Modeling of future electricity generation and emissions assessment for Pakistan. *Processes, 7*(212), 1–26.

Moxnes, E. (1990). Inter-fuel substitution in OECD—European electricity production. *System Dynamics Review, 6*(1), 44–65.

Naill, F. (1992). A system dynamics model for national energy policy planning. *System Dynamics Review, 8*(1), 1–19.

Nasirov, N., O'Ryan, R., & Osorio, H. (2020). Decarbonization tradeoffs: A dynamic general equilibrium modeling analysis for the Chilean power sector. *Sustainability, 12,* 8248.

Naz, S., Sultan, R., Zaman, K., et al. (2019). Moderating and mediating role of renewable energy consumption, FDI inflows, and economic growth on carbon dioxide emissions: Evidence from the robust least square estimator. *Environmental Science and Pollution Research, 26,* 2806–2819.

Panda, D. (2011). Impact of renewable energy sources in the power supply of India—A system dynamics approach. *International Journal of Power System Operation and Energy Management, 2*(3, 4), 2231–4407.

Pandey, K., & Rastogi, H. (2019). Effect of energy consumption & economic growth on environmental degradation in India: A time series modeling. *Energy Procedia, 158,* 4232–4237.

Park, Y., Ahn, S., Yoon, B., Koh, H., & Bunn, W. (2007). Investment incentives in the Korean electricity market. *Energy Policy, 35*(11), 5819–5828.

Qudrat-Ullah, H. (2013). Understanding the dynamics of electricity generation capacity in Canada: A system dynamics approach. *Energy, 59*, 285–294.

Qudrat-Ullah, H. (2015a). Independent power (or pollution) producers? Electricity reforms and IPPs in Pakistan. *Energy, 83*(1), 240–251.

Qudrat-Ullah, H. (2015b). Modeling and simulation in service of energy policy. *Energy Procedia, 75*, 2819–2825.

Qudrat-Ullah, H. (2017). How to enhance the future use of energy policy simulation models through ex-post validation. *Energy, 120*(1), 48–66.

Qudrat-Ullah, H. (2022). A review and analysis of renewable energy policies and CO_2 emissions of Pakistan. *Energy, 238*, 121849.

Qudrat-Ullah, H., Akrofi, M., & Kayal, A. (2020). Analyzing actors' engagement in sustainable energy planning at the local level in Ghana: An empirical study. *Energies, 13*(2020), 1–29.

Raggad, B. (2018). Carbon dioxide emissions, economic growth, energy use, and urbanization in Saudi Arabia: Evidence from the ARDL approach and impulse saturation break tests. *Environmental Science and Pollution Research, 25*(15), 14882–14898.

Rashid, M. I., Benhelal, E., & Rafiq, S. (2020). Reduction of greenhouse gas emissions from gas, oil, and coal power plants in pakistan by carbon capture and storage (CCS): A review. *Chem Eng Rev, 43*(11), 2140–2148.

Raza, M. Y., & Lin, B. (2020). Decoupling and mitigation potential analysis of CO_2 emissions from Pakistan's transport sector. *Science of the Total Environment, 30*, 139000.

Rehman, U., Cai, Y., Siyal, Z. A., Mirjat, N. H., Fazal, R., & Kashif, S. U. R. (2020). Cleaner and sustainable energy production in Pakistan: Lessons learnt from the Pak-TIMES Model. *Energies, 13*(108), 1–21.

Rue du Can, S., Khandekar, A., Abhyankar, N., Phadke, A., Khanna, N., Fridley, D., & Zhou, N. (2019). Modeling India's energy future using a bottom-up approach. *Applied Energy, 238*, 1108–1125.

Samargandi, N. (2017). Sector value addition, technology, and CO_2 emissions in Saudi Arabia. *Renewable and Sustainable Energy Reviews, 78*, 868–877.

Saysel, K., & Hekimoglu, M. (2013). Exploring the options for carbon dioxide mitigation in Turkish electric power industry: System dynamics approach. *Energy Policy, 60*, 675–686.

Shafiei, E., Davidsdottir, B., Leaver, J., Stefansson, H., & Asgeirsson, E. I. (2015). Simulation of alternative fuel markets using integrated system dynamics model of the energy system. *Procedia Computer Science, 51*, 513–521.

Shahzada, M., Nawaz, N., & Alvi, S. (2021). Energy security for socio-economic and environmental sustainability in Pakistan. *Heliyon, 4*, e00854. https://www.cell.com/action/showPdf?pii=S2405-8440%2818%2932303-X. Accessed 20 Jan 2021.

Shearer, C., Fofrich, R., & Davis, J. (2017). Future CO_2 emissions and electricity generation from proposed coal-fired power plants in India: Proposed coal plants in India. *Earth's Future, 5*, 408–446.

Sisodia, G. S., Sahay, M., & Singh, P. (2016). System dynamics methodology for the energy demand fulfillment in India: A preliminary study. *Energy Procedia, 95*, 429–434.

Sterman, J. (2000). *Business dynamics: Systems thinking and modeling for a complex world*. McGraw-Hill.

Tiewsoh, S., Jirásek, J., & Sivek, M. (2019). Electricity generation in India: Present state, future outlook and policy implications. *Energies, 12*, 1361. https://doi.org/10.3390/en12071361

Uddin, R., Shaikh, A. J., Khan, H. R., Shirazi, M. A., Rashid, A., & Qazi, S. A. (2021). Renewable energy perspectives of Pakistan and Turkey: Current analysis and policy recommendations. *Sustainability, 13*, 3349. https://doi.org/10.3390/su13063349

Usman, A., Ullah, S., Ozturk, I., Chishti, M. Z., & Zafar, S. M. (2020). Analysis of asymmetries in the nexus 769 among clean energy and environmental quality in Pakistan. *Environmental Science and Pollution Research, 1–12*(771), 63.

Chapter 5
Modeling Methodology for Energy Policy

5.1 Introduction

Although various modeling approaches and models (e.g., econometric, optimization, input–output models, LEAP, MARKAL GAINS) have served the energy domain well (Qudrat-Ullah, 2015; Rehman et al., 2020), system dynamics methodology stands out due to its ability to capture the essential feedback structures, which are responsible for generating outcomes for any energy system (Qudrat-Ullah, 2022; Sterman, 2000). Specifically, system dynamics' models explicitly allow to:

i. Account for both physical (e.g., in construction of power plants and regulatory approvals) and information delays (e.g., in the approval of licenses and permits from various governmental agencies),
ii. Represent the non-linear relationships among various variables of the energy system (e.g., increasing electricity tariffs, after a while, will lead to relatively different sensitivity levels for industrial and residential users),
iii. Effectively deal with soft variables (e.g., the morale and satisfaction level of power plant operators),
iv. Distinguish between the desired state (e.g., desired level of CO_2) and 'actual' state of variables (e.g., the actual level of CO_2 based on the existing supply mix of a country),
v. Account for the dynamic interactions among the energy system sectors and the variable of these sectors,
vi. Link micro-level decision making (e.g., investments in renewable and cleaner electricity generation capacity) with macro-level outcomes structures (e.g., electricity supply mix and related CO_2 emissions),
vii. Endogenous determination of energy demand, which is often assumed exogenous in traditional modeling approaches,
viii. To build the framework for the transient behavior (e.g., adjustments for new energy production technologies in response to resource price dynamics typically create transient behavior) (Qudrat-Ullah, 2013),

H. Qudrat-Ullah, *Understanding the Dynamics of Nuclear Power and the Reduction of CO_2 Emissions*, https://doi.org/10.1007/978-3-031-04341-3_5

ix. Identify the energy policy levers and investigate the dynamic implications of energy policies directed at such policy levers (Muller et al., 2013),
x. Analyze multidimensional impacts of policy implementation and identify potential "double and triple dividends" intervention (Bassi et al., 2013),
xi. Explicitly model the perceptions of people (e.g. regarding the adoption of generation technologies (Qudrat-Ullah et al., 2020), and
xii. Capturing the endogenous complexity of energy systems, which is considered responsible for the behavior of the system (Al-Sarihi & Bello, 2020; Qudrat-Ullah, 2015, 2022; Sterman, 2000).

These modeling strengths of System Dynamics (SD) (Forrester, 1961) are a major motivation for us to design, develop, validate, and apply an SD model for the policy assessment and design of the energy policy of Pakistan. It should be noted that the basic premise of system dynamics methodology is: "structure of a system (e.g., feedback loops, delays, and non-linear relationships among the variables of an energy system) drives its performance (e.g., the evolution of electricity-related CO_2 emissions)" (Qudrat-Ullah, 2016; Sterman, 2000). If the decision-makers understand the structures of the energy systems well, it is hoped, they will make better decisions—achieve their objectives (e.g., reduction of CO_2 emission). While other models (e.g., optimization model) are prescriptive or heavily dependent on past data (e.g., econometric models), system dynamics simulation models are descriptive and aim at the identification of causal relationships among variables that are responsible for the outcome behavior of the systems (Bassi et al., 2013; Qudrat-Ullah, 2016). Therefore, in this chapter, we present a detailed description of our SD model, MDESRAP. In the following chapter, we will demonstrate how did we validate MDESRAP to be used for Pakistan's energy policy assessment and alternate policy design.

5.2 The Development of SD Model, MDESRAP

Although the genesis of MDESRAP, the SD-based model, was the evaluation of Pakistan's energy policy in 2001 (Qudrat-Ullah & Davidsen, 2001), for this project we have revised it substantially. Therefore, it is worth describing MDESRAP's key structural elements here. Overall, MDESRAP comprises seven sectors of the electricity supply system of Pakistan:

1. Electricity Demand Sector
2. Investment Sector
3. Capital Sector
4. Resource Sector
5. Production Sector
6. Environment Sector
7. Costs and Pricing Sector

Figure 5.1 presents an integrated overview of the sectors of MDESRAP. The first

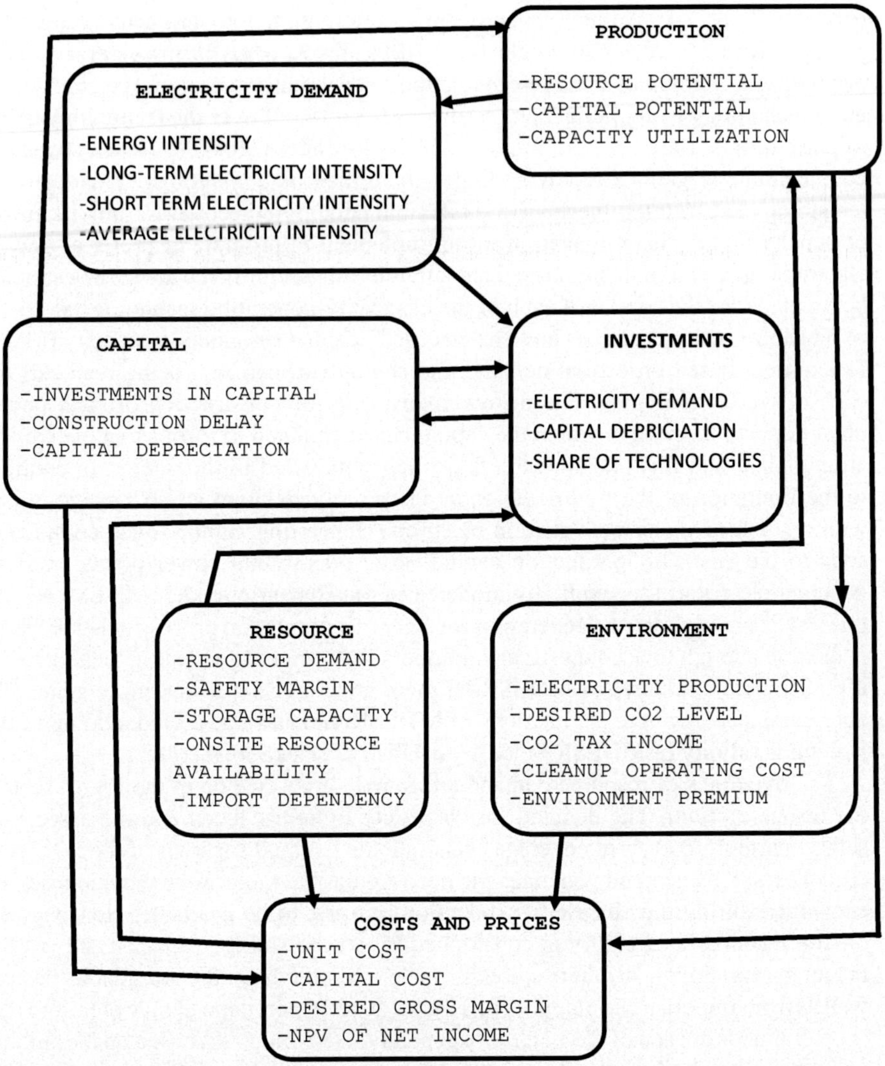

Fig. 5.1 Overview of the SD model (i.e., MDESRAP) sectors

sector describes how the electricity demand (MWh/Year) is generated, based on GNP and the electricity intensity of GDP. GDP is exogenous to the model. The average electricity intensity is determined based on the average price of electricity while also taking into account the changes (appreciation/ depreciation) in electricity-generating capital.

The second sector describes how the investments in electricity capital are made across the generating technologies (coal, oil, gas, solar, wind, nuclear, and hydropower plants), based on the costs of these technologies, environment premium (due to CO_2 tax), investment incentive premium (e.g., the government's long-term power purchase agreements might favor IIPPs' investments in low-operating cost generating technologies), and resource import dependency premium (i.e., the generation technology using indigenous resources is preferred over the technology using imported-fuel). Based on the multinomial logit model (Moxnes, 1990), the least-cost technology gains a relatively larger share than its competitors. This leat cost technology substitution mechanism is flexible enough to account for any incentive or penalty that decision-makers want to implement to promote or prefer a specific electricity generation technology. The output of this sector is the actual investments being made for the capital of each of the electricity-generating technologies.

The third sector describes how the electricity capital (production capacity [MW]) is acquired, based on capital demand and supply (investments being realized) for each of the technologies. The delays involved in the construction of each power plant impact the speed at which the capital is accumulated to form a complete operational plant and this phenomenon is explicitly modeled in this sector. In contract to the limitation of the traditional econometric models, here a two-vintage model with a stock of under construction electricity-generating capital, once completed, adds to the stock of 'production capital' (i.e., operational power plants-ready to generate electricity), is explicitly modeled in our dynamic model, MDESRAP. As this tow-stage capital (i.e., generation capacity) accumulation process in MDESRAP is flexible enough to account for any number of electricity-generating technologies, it makes such structures of MDESRAP more generic. As a result, more generalizable outcomes from the applications of MDESRAP can be expected, adding to the structural validity of MDESRAP (Qudrat-Ullah & BaekSeo, 2010).

The dynamics of resource demand and supply are modeled in the fourth sector; the resource sector. The demand for electricity to be produced and the associated safety margin (short-term coverage) generate the demand for each of the resources (Btu/Year). This demand generates the need for on-site fuel resource, which is to be consumed during the electricity production process, to be available. However, the on-site resource availability is constrained by (i) the storage capacity and (ii) the resource availability. A fule resource is made available from the indigenous base as well as from imports. The output of this sector is the import dependency premium that is determined for each of the electricity generation technologies, based on the amount of fuel imported. More the imports of a fule resource will make it a less attractive electricity generation technology. On the other hand, if a generation technology uses domestic fuel resources (i.e., no imported fuel is used in electricity generation) then it is favored with a positive premium known as the "import dependency" premium.

The electricity production based on the resource potential and the production capacity capital is described in the fifth sector, the production sector. It should be noted that the availability of the resource (i.e., fuel) at the site of the power plant(s) is needed to begin the operation of a fuel-based power plant. Availability of domestic resources ver imported fuel will affect how an electricity generation technology is gaining or losing its share in the overall electricity supply mix of the country. Likewise, production capacity capital here refers to the "operational" capacity of electricity-generating technologies or power plants. A maximum of both the production capacity (i.e., in MW) and the resource potential (i.e., in MW) will determine how much actual electricity (MWh) is produced.

The sixth sector describes how the environment premium ($ / MWh) is determined for each of the electricity-generating technologies, based on the amount of electricity produced, the CO_2 emission intensity of fuel, and CO_2 tax rates, while taking into account the reference/desired CO_2 emission limit. However, CO_2 emissions can travel and can cross the boundary of the producing country. In our formulation, we have accounted for the total CO_2 produced, how much of these emissions were treated, and how much of these CO_2 emissions escaped into the boundaries of other neighboring countries. This accounting for all electricity-related CO_2 is critical in determining the associated or potential clean-up costs. These additional clean-up costs can be used to determine the appropriate range of CO_2 tax for the respective generation technologies. That, in turn, will determine the increasing share of the least-costing technologies by adding a supporting premium to renewable and nuclear power.

The last sector describes financial conditions, such as costs, prices, and net income as they develop as a result of investments. The following section describes the sectorial modeling and overview of MDESRAP with a complete set of mathematical equations, as were coded in POWERSIM™, for each of its sectors.

5.3 The Sectorial Modeling and Overview of MDESRAP

In the description of each sector/sub-sector, we will first describe the explanation of sectorial interactions followed by the listing of corresponding mathematical equations of the system relationships inboxes.

5.3.1 Electricity Demand Sector

Electricity demand is one of the major uncertainties of an electricity supply system. Researchers have used a variety of models to represent or model electricity demand. A recent review by Verwiebe et al. (2021) analyzed 416 articles published from 2015 to 2020 and found that among the five commonly used modeling techniques: statistical, machine learning, metaheuristic, stochastic/fuzzy/grey, and engineering-based techniques, the engineering-based techniques, which include simulation modeling approach that we are using here in this book, is the most used modeling technique by energy domain researchers. In the demand sector of our dynamic model, MDESRAP, the total electricity demand is being generated, based on the exogenously given GDP (of Pakistan), and the average electricity. An increase in the average electricity intensity results in more electricity vice versa. The better economic conditions (increased GDP) also increase the demand for electricity supply. The mathematical equation for the total electricity demand is given in Box # 5.1.

Box # 5.1: Mathematical Equation for Electricity Demand

```
TotElecDemand = ((AvgElecIntensity)*GDP_of_Pakistan)
```

5.3.1.1 Average Electricity Intensity

The average electricity intensity (AvgElecIntensity) is represented by a co-flow structure as depicted in Fig. 5.2. In this co-flow structure, two stocks, (i) production capital and (ii) electricity intensity move in proportion in both "under construction" and "operational" phases. If one stock increases (e.g., the capital under construction), its corresponding co-flow stock (i.e., under construction electricity intensity) also increases and vice versa. In other words, this co-flow structure is influenced by the physics of the electricity generating capital. Accordingly, the average electricity intensity increases when the capital is increased and it decreases when the capital depreciates.

The input to the electricity intensity stream is the long-term indicated electricity intensity (LongTIndcElecIntensity) while the capital under construction (ElecCapt-Cons) is the input to the capital. The reference electricity intensity and the effect

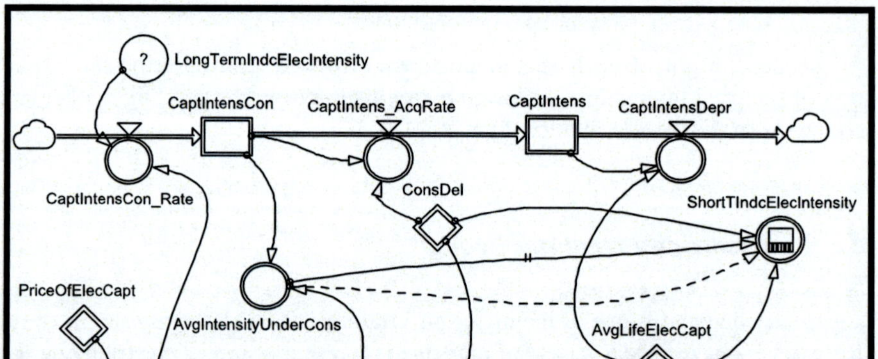

Fig. 5.2 Co-flow structure of average electricity intensity

of electricity price on electricity intensity determine the long-term indicated electricity intensity. The reference electricity intensity represents our case-specific base year (2008) value. The effect of electricity price on electricity intensity is dependent upon (i) the effect of electricity price on demand, (ii) the average electricity price, and (iii) the reference electricity price. If the price of electricity decreases (relative to the reference electricity price), the electricity intensity exhibits a growth pattern. Conversely, electricity consumption declines when the price rises. The corresponding mathematical equations are listed in Box # 5.2:

Box # 5.2: Mathematical Equations for Average Electricity intensity

```
AvgElecIntensity = REfElecIntensity
AvgElecIntensity = +dt*ChngInElecIntensity
ChngInElecIntensity=(ShortTIndcElecIntensity-
AvgElecIntensity)/TimeToRealizeChng
ShortTIndcElecIntensity=IF(ARRSUM(ElecCapital)>0,
ARRSUM(ElecIC)/ARRSUM(ElecCapital),0)
ElecIC = LongTermIndcElecIntensity*ElecCapital
ElecIC = +dt*ElecIC_Rate -dt*ElecIDec
ElecICon_Rate = LongTermIndcElecIntensity*(ElecCaptCons)
ElecIC_Rate = ElecICon/ConsDel
ElecIDec = ElecIC/AvgLifeElecCapt
ElecICon = ElecCaptUndCons*LongTermIndcElecIntensity
ElecICon = -dt*ElecIC_Rate + dt*ElecICon_Rate
ElecICon_Rate = LongTermIndcElecIntensity*(ElecCaptCons)
LongTIndcElecIntensity          =          REfElecInten-
sity*EffectOfPriceOnElecIntensity
EffectOfPriceOnElecIntensity=EXP(EffectOfElecPriceOnDemand*
LN(AvgPriceOfElec/RefElecPrice))
RefElecPrice = GRAPHCURVE(TIME,1980,5,[5.9,13.3,18.2,19.7
"Min:1;Max:50"])
EffectOfElecPriceOnDemand = -.02
REfElecIntensity = .50
```

5.3.1.2 Average Price of Energy

The shares (in total energy consumption) and the prices of electricity and its competitive alternative (oil in our case) determine the average price of energy. The price of the substitute (PriceofOther) is assumed a linear extrapolation of the historic price of oil. While the electricity price is endogenously calculated in the model as the average price of electricity (AvgElecPrice). The mathematical Equation for the Average Price of Electricity is listed in Box # 5.3.

Box # 5.3: Mathematical Equations for Average Price of Electricity

```
AvgPriceOfEnrgy = (AvgPriceOfElec*SharOfElec) + (PriceO-
fOther*ShareOther)
AvgPriceOfElec = SUM(t = Coal..Gas;ElecProdRate(t)*
PriceOfElec(t))/ARRSUM(ElecProdRate)
```

5.3.1.3 Share of Electricity and Oil

The share of electricity consumption (ShareElec) is determined through a substitution mechanism, represented in the above equations. The average price of electricity sends the signal to the market calling for investments in electricity-generating capital and oil exploration/production capital. If this reference electricity price for investments (RefElecPriceforInvestmentinOther) is relatively high compared to the price of oil and is sustained for a certain period of observation, then the indicated share of production capacity of oil (IndShareOfProdCapacityOther) is increased. The period of observation will filter out the seasonal/sudden price changes. This relationship (between the sustained relative oil price and the indicated share of production capacity of oil) is non-linear, as exhibited in Fig. 5.3.

This relationship is identified based on the historical data (Pakistan's case). The extreme points of the graph represent the limit of substitution. For some consumers, depending upon the application, it may be impossible to switch over despite the significantly more attractive price of the substitute. For example, the price of electricity may become much cheaper (e.g., more than six times) than the price of oil, but certain consuming industries (e.g., transport sector) will still be constrained to substitute for electricity. Similarly, certain consuming appliances (e.g., hospital equipment) will be dependent upon electricity regardless of how inexpensive the substitute (oil) is.

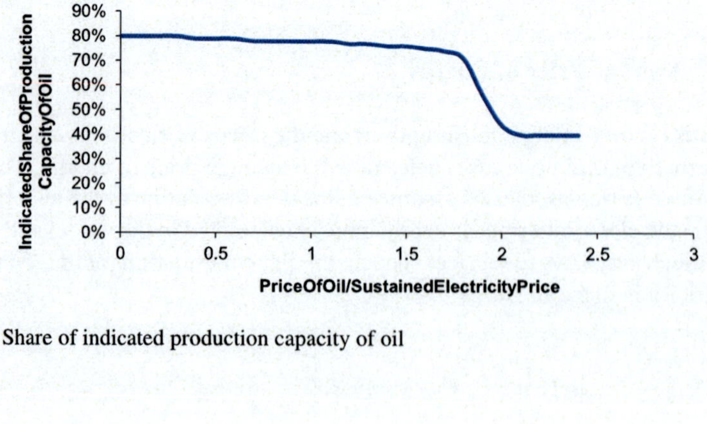

Fig. 5.3 Share of indicated production capacity of oil

This indicated share of oil together with total energy demand (TotalEnergyDemnd) determines the indicated production capacity of oil (IndProdCapOther). The indicated production capacity of oil determines how much new capacity is to be added. The production capacity of oil (ProdCapacityOfOther) is dependent upon the indicated production capacity and the capacity adjustment time (ProdCapacityAdjTime). The capacity adjustment time determines the speed with which the capacity is added. If the indicated share of oil is (i.e., to a dominant range greater than 70%) increasing and the price of electricity is relatively high, then the production capacity of oil is added quickly (3 years in our case).

The production of the substitute (ProdOfOther) depends on the production capacity and the capacity utilization, as is described by the above equations. The indicated production of the substitute corresponds to the production justified by the current price of electricity, taking capital as well as the variable costs into consideration. If the production indicated this way is below the capacity, the production of the substitute is gradually shut down. But even if the price of electricity falls to indicate a capacity reduction, the price may still cover variable costs as well as a fraction of the past capital investments. Assuming a different cost structure, the impact of certain price falls will vary: if the price of electricity is relatively close to the price of oil, a large portion of the industry may still cover most of its investments and the capacity utilization will therefore be relatively higher. If, on the other hand, the price of electricity is considerably lower than the price of oil, the price fall will cause correspondingly larger portions of the industry to attain the break-even point. In this case, the variable costs may not be recovered and a shutdown is indicated. This variable sensitivity to a certain fall in relative price is presented in the table describing the capacity utilization, depicted in Fig. 5.4. This structure is adopted from the work of Pål Davidsen (1986, Section 3.1.9). As is illustrated in the next chapter, the use of the existing model structure adds to the structural validity of a model (Qudrat-Ullah, 2008; Qudrat-Ullah & BaekSeo, 2010)—a right structure of a model for the right reasons (Barlas, 1989).

The share of the substitute (ShareOther) in the total energy consumption is determined by the actual production of oil (ProdOfOther). The larger the production, the larger is its share of energy consumption. The increased total consumption also results in an increased share of energy consumption. This share of the substitute's consumption determines how much electricity is consumed (ShareElec).

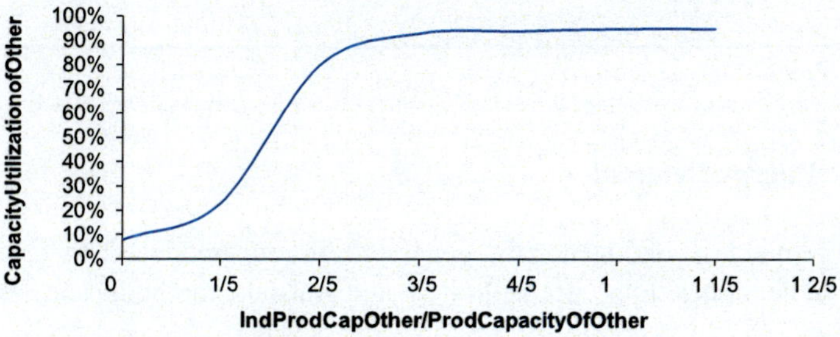

Fig. 5.4 Capacity utilization of oil production

Box # 5.4: Mathematical Equations for Share of Electricity and Oil

```
SharOfElec = 1-ShareOther;ShareOther =
IF(TotalEnergyDemnd>0,(ProdOfOther/
TotalEnergyDemnd),0); ProdOfOther = ProdCapacity-
OfOther*CapUtilOfOther
ProdCapacityOfOther = DELAYINF(IndProdCapOther, ProdCapac-
ityAdjTime,3)
IndProdCapOther = IndShareOfProdCapaci-
tyOther*TotalEnergyDemnd
IndShareOfProdCapacityOther=GRAPHCURVE(PriceOfOther/
RefElecPriceforInvestmentinOther,0,0.5,[0.94,0.94,0.94,0.94,
0.94,0.93,0.92,0.91,0.89,0.87,0.84,0.7,0.7,0.7
"Min:0;Max:1"]);PriceOfOther=GRAPHCURVE
(TIME,1980,5,[19.8,22.6,28.8,33.7,40.2,44.2,
45.8,45.8,48.7,50.4,50.9"Min:0;Max:60"])
RefElecPriceforInvestmentinOther=
DELAYINF(AvgPriceOfElec,TimeToEstRefElecPrice,3)
ProdCapacityAdjTime=IF(IndShareOfProdCapacityOther>=.70,(IF
(SustainedRefElecPrice(1)>=1.5*PriceOfOther,3,4)),6)
SustainedRefElecPrice = IF(difference=0,
RefElecPriceforInvestmentinOther,0.00001)
difference=IF((ABS(RefElecPriceforInvestmentinOther-
DelayedRelativeChange))
<=.01,0,(RefElecPriceforInvestmentinOther-
DelayedRelativeChange))
DelayedRelativeChange = DELAYPPL
(RefElecPriceforInvestmentinOther, TimeToObserveSustained-
Change);
CapUtilOfOther=GRAPHCURVE(IndProdCapOther/ProdCapacity
OfOther,0,0.2, [0.08,0.23,0.8,0.93,0.94,0.95,0.95
"Min:0;Max:1″]);TimeToEstRefElecPrice=;
ElecIDec = ElecIC/AvgLifeElecCapt;ElecICon = ElecCaptUnd-
Cons*LongTermIndcElecIntensity;
ElecICon=-dt*ElecIC_Rate+dt*ElecICon_Rate;ElecICon_Rate=
LongTermIndcElecIntensity*(ElecCaptCons);LongTIndc
ElecIntensity=EfElecIntensity*EffectOfPriceOnElecIntensity;
EffectOfPriceOnElecIntensity=EXP(EffectOfElecPrice
OnDemand*LN(AvgPriceOfElec/RefElecPrice));RefElecPrice=
GRAPHCURVE(TIME,1980,5,[5.9,13.3,18.2,19.7
"Min:1;Max:50″]); EffectOfElecPriceOnDemand= -.02;
REfElecIntensity = .50;
```

5.3.1.4 Energy Demand

Energy demand is endogenously generated in our model. This estimation of energy demand has long been studied and various methods and formulations have been proposed (Altaf, 2018; Akbar et al., 2017). The energy demand (**TotalEnergyDemnd**) is dependent on the exogenously given GDP[1] (of Pakistan), and the corresponding energy intensity (EnrgyIntensity). An increase in the energy intensity results in higher energy demand while a decrease causes a decline. In addition, the growth in GDP requires more energy to support its commercial (and industrial) activities. The reference energy intensity (RefEnrgyIntensity) and the effect of energy price on energy intensity (EffOfPriceOnEnrgyIntensity) determine the indicated energy intensity of GDP (IndcEnrgyIntensity). The reference energy intensity represents our case-specific base year (2008) value. The effect of price (AvgPriceOfEnrgy) on energy intensity varies: a price increase leads to a reduced energy consumption regime while the decrease in the price of energy allows for a more intense consumption of energy. The tendency to adjust investments to a change in the price of energy is determined by the price elasticity of energy demand, i.e. the effect of price on demand (EffectOfPriceOnDemand). This represents a non-linear relationship between the price of electricity and demand for electricity. The price elasticity is referring to a reference energy price (RefEnrgyPrice). The consequence of a downward capital adjustment, concerning the energy intensity, is delayed by an amount corresponding to the physical life of the electricity generating capital (AvgLifeOfElecCapt) on average. This reflects the time required to renew the capital stock (i.e., to add new capacity to replace the depreciated capacity). The equations for the Total Energy Demand are given in Box # 5.5.

Box # 5.5: Mathematical Equations for Total Energy Demand

```
TotalEnergyDemnd = GDP_of_Pakistan*EnrgyIntensity*1,000,000
EnrgyIntensity = RefEnrgyIntensity
EnrgyIntensity = +dt*ChngEnrgyIntensity
ChngEnrgyIntensity = (IndcEnrgyIntensity-
EnrgyIntensity)/TimeAdjEnrgyIntensity
IndcEnrgyIntensity = EffOfPriceOnEnrgyInten-
sity*(RefEnrgyIntensity)
EffOfPriceOnEnrgyIntensity=EXP(EffectOfPriceOnDemand*
LN(AvgPriceOfEnrgy/RefEnrgyPrice))
RefEnrgyPrice=
GRAPHCURVE(TIME,1980,1,[15.8,17.6,18.2,20,
20,21.3,22.9,25.5"Min:0;Max:60"])
RefEnrgyIntensity = 0.53
EffectOfEnrgyPriceOnDemand = -0.04
```

[1] Our dynamic model, MDESRAP, utilized the GDP data of Pakistan and then use extrapolation method for the future years of this i.e., from 2008 to 2021 (existing data) and from 2022 to 2035 (extraplolated data).

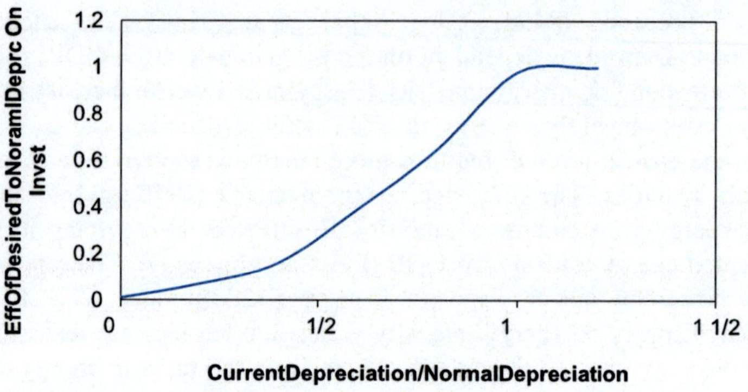

Fig. 5.5 Effect of desired to normal depreciation on investment

5.3.2 Investment Sector

Most governments attempt to attract private investments in energy sector projects through various policy incentives and initiatives. In our model, IPP[2]s investments in each of the electricity generating capitals (IPPInvst) are based on the indicated investments in electricity-generating capital (IndicInvstInElecCapt). The total demand of electricity-generating capital (TotDesElecCapt) and the share (ActualShareTech) of each of the competing technologies (that is coal, gas, oil, and hydro-based power plants) determine the indicated investments. The capital investments are made to make up the discrepancy between the existing capital (ElecCapital) and the new demand for that capital (ElecCaptDemand) as well as to replace the depreciated capital (ElecCaptDepr). The effect of current to desired depreciation on investments is considered, as depicted in Fig. 5.5. When there is no desired depreciation, the capital demand includes the replacement for the normal depreciation. However, if capital shut down is initiated, only a fraction of normal depreciation is replaced/ordered, as is evident from the following graph.

An electricity-generating technology (power plant) obtains a higher market share if the associated total cost for electricity production is relatively lower, as is explained in the following section. Here, the dynamic substitution of renewable and non-renewable electricity-generation technologies, as is suggested by some earlier studies (Rashid et al., 2020; Uddin et al., 2021; Valasai et al., 2017), takes place. The mathematical equations for IPP Investment Function are given in Box # 5.6.

[2] IPPs: Independent Power Producers.

Box # 5.6: Mathematical Equations for IPP Investment Function

```
IPPInvst = IndicInvstInElecCapt
IPPInvst = +dt*InvstRate
InvstRate= ((IndicInvstInElecCapt-IPPInvst)/AdjTimeForInvst)
IndicInvstInElecCapt = TotDesElecCapt*PriceOfElecCapt
*ShareOfTechTotDesElecCapt=
IF(DiscInElecCapt=0,ARRSUM(ElecCaptDepr),IF(DiscInElecCapt
<0,EffOfDesiredToNoramlDeprcOnInvst*ARRSUM(ElecCaptDepr),
ARRSUM(ElecCapt Depr)+ ElecCaptCorrection)))
DiscInElecCapt = ElecCaptDemand-ARRSUM(ElecCapital)
ElecCaptDemand = (TotElecDemand/(ElecCaptProdctivity))
EffOfDesiredToNoramlDeprcOnInvst=GRAPHCURVE
(DesiredDepreciation/                NormalDepreciation,-
3,0.5,[0.01,0.07,0.16,0.39,0.65,0.95,1
"Min:0;Max:1"])
DesiredDepreciation = DiscInElecCapt/TimeToAdjstElecCapt
NormalDepreciation = ARRSUM(ElecCaptDepr)
ElecCaptCorrection = IF(DiscInElecCapt>0,
DiscInElecCapt/TimeToAdjstElecCapt,0)
AdjTimeForInvst = 2
TimeToAdjstElecCapt = 6
```

5.3.2.1 Share of Each Electricity Generating Technology

The share of each competing electricity-generating technology (ShareTech) in the new capital investments being made is represented by a lagged version of the planned shares (PlannedShareTech). The average delay time (ConsTime) represents the construction time of new power plants. Thus, we assume that the choice of electricity-generating technology cannot be altered after the plant construction has begun.

Numerous factors are likely to influence the share planned for each technology when the investments decision is being made (Moxnes, 1990). However, we focus, in addition to the standard cost elements, on the investment incentive premium (InvestmentIncentivePrem), the import dependency premium (ImportDependencyPremium), the environment premium (EnvPrem) for each technology, and we lump the rest in a power plant technology choice premium (OtherPrem). The standard cost elements includes capital costs (CaptCost), operating costs (OprtCost), and fuel costs (FuelCost). Annualized capital cost for each technology is determined based on the capital unit cost (CpatUnitCost) and the average (physical) life of the capital (AvgLifeElecCapt). Capital unit costs vary due to different price offers by a multitude of capital equipment suppliers (Moxnes, 1990). The operating costs include other than fuel costs. Fuel costs are normalized by the respective fuel efficiencies. These cost elements are described in the cost and pricing sector of the model.

Once the total cost for each technology (TotCostElecTech) is determined, we apply a multinomial logit (MNL) model to obtain the planned share of each technology (PlannedShareTech). We have built this modified total cost structure, with the inclusion of the three kinds of premiums, based upon the work of Erling Moxnes (Moxnes, 1990).

The MNL assumes that the individual electricity producer chooses the cheapest alternative. In this model, the market share of the electricity generating technology is gradually decreased when the total cost of that technology is increased relative to competing technologies. The fact that different IPPs face different total costs (for example due to different equipment suppliers) implies that IPPs was taken as a whole behave differently than each IPP. For example, not all the IPPs will choose gas or hydro or coal-based power plants even though the average costs of oil-fired power plants are slightly above the average costs of gas or hydro or coal-based power plants. Some of the IPPs will still find oil-fired power plants to be the least expensive choice. This represents a more flexible approach to the modeling of the substitution processes than that of a much more rigid approach where an electricity-generating technology is assumed either competitive or not (Dyner, 2000; Gholizad et al., 2017).

The MN[3]L has only one parameter α (distribution parameter); except for the parameters that apply to the cost function. When α takes an extreme value, then the aggregate choice of the whole sector (all the IPPs) for a technology mimics the choice of an individual IPP. The lower values of α represent greater variation between the individual choices and the aggregate (sector) choice. When the total costs of all the technologies are equal, the market share is split into equal shares. Also, the (own) price elasticity of a technology ($\in_{ii}$) is variable. It depends upon (i) share planned for each technology (PlannedShareTech), fuel price (ResPrice), and fuel efficiency (FuelEfficiency): $\in_{ii} = -\alpha$ (1-PlannedShareTech) * (ResPrice) / (FuelEfficiency) and $\in_{ij} = \alpha$ (PlannedShareTech) * (ResPrice) / (FuelEfficiency) for cross-price elasticity, as shown by Moxnes (1990). The mathematical equations for the Share of generation Technologies are listed in Box # 5.7.

Box # 5.7: Mathematical Equations for The Share of generation Technologies

```
ShareOfTech = [.047,13.12,.27,9.19]
ShareOfTech = +dt*TechConstRate
TechConstRate = DiscInShareOfTech/ConsDel
DiscInShareOfTech = (PlannedShareOfEachTechnology-
ShareOfTech)
PlannedShareOfEachTechnology=EnrgyTechShare(t)/
ARRSUM(EnrgyTechShare)
EnrgyTechShare = EXP(-DistPara(t)*CostOfElecTech(t))
CostOfElecTech=(((CaptCostOfElecProd))+Fuel_Cost
```

[3] MNL: Multinomial Logit model.

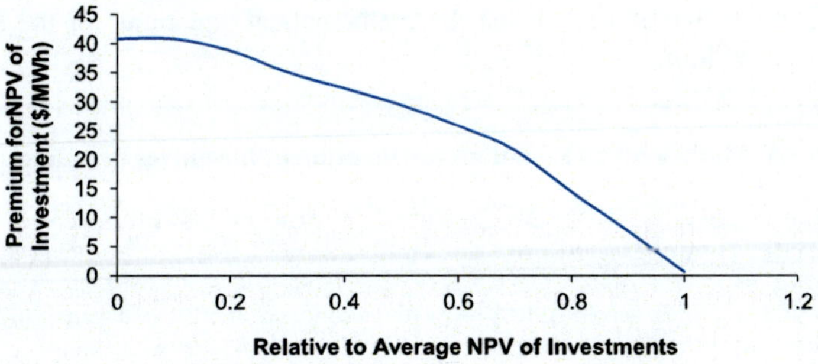

Fig. 5.6 NPV of investment premium

```
+(OpertCost*10)+(Investment
IncentivePrem)+OtherPrem+(ImportDependencyPremium)+(EnvPrem
))/10DistPara = 0.02488
```

5.3.2.2 Investment Incentives Premium

The investment premium (InvestmentIncentivePrem) reflects the impact of policy incentives on the investments being made for electricity-generating technologies. The estimation of this relationship arises as one of the results of our work. This identified premium seems reasonable. Two types of policy incentives are included, namely the net present value and power (electricity) purchase guarantee. If a technology generates an earlier income stream, that will have a positive (a higher) impact on the investment decision of IPPs. An IPP will preferably invest in this electricity-generating technology. Similarly, if the govt. guarantees to purchase all the electricity from the supplier, then the technology with low operating costs might attract more of the IPPs investments. This assertion is based on the assumption that, given the guaranteed power purchase agreements, the IPPs will take a long-term perspective on the investments. The weighted average of the effects of both the policy incentives[4] is used to compute the effect of incentives on IPPs investments. The higher the effect of an incentive, the lower is the premium. That reflects an attraction of investment in that technology by the IPPs. The NPV[5] of investment premium is represented by a table function, as shown in Fig. 5.6.

The maximum premium, represented in the table function, is of value equal to 40.80 $ / MWh. This represents the case when the investor sees this investment opportunity as of very low-income potential and is least interested in making the

[4] Other policy incentives like the recent introduction and promotion pf netmetring (Ziras et al., 2021), are modeled separately in our dynamic model MDESRAP.

[5] NPV: Net Present Value.

investments. In Box # 5.8, we list the mathematical equations for the Investment Incentives Premium.

Box # 5.8: Mathematical Equations for Investment Incentives Premium

```
InvestmentIncentivePrem=NPVOfInvstPrem*Weightage
ForNPVIncentives+CaptCostInvstPrem*  WeightageForCapital-
Cost
NPVOfInvstPrem=GRAPHCURVE(RelativetoAvgNPVOfInvst/ARRSUM
(RelativetoAvgNPVOfInvst),0,0.1,[40.8,39.6,39.1,
38.8,38.5,36.8,34.8,30,25.5,17.5,5.7"Min:0;Max:60"])
CaptCostInvstPrem=GRAPHCURVE(RelativetoAvgCapitalCost/ARRMAX
(RelativetoAvgCapitalCost),0,0.1,[0,0,1.1,
4.8,8.5,13.6,21.5,29.4,36.8,45.6,49.8"Min:0;Max:60"])
RelativetoAvgCapitalCost =CaptCostOfElecProd/AvgCapitalCost
RelativetoAvgNPVOfInvst=NPVOfInvst/AverageNPVOfInvest
WeightageForNPVIncentives =1-WeightageForCapitalCost
WeightageForCapitalCost= .8
```

5.3.3 *Electricity Capital Sector*

In the model, the electricity (generating) capital (ElecCapt) means the physical machinery, buildings, equipment, etc. used to produce electricity from raw fuel (resource). It is measured in the form of production capacity (i.e., MW[6] capacity). The formulation represented in the above equations represents a kind of generic structure of the capital stock. The electricity-generating capital stock (ElecCapt) is increased by the capital acquisition rate (ElecCaptAcqRate) and decreased by the capital depreciation rate (ElecCaptDeprRate). Capital is initialized at its equilibrium level (i.e., the existing production capacity of Pakistan in 2008). Capital depreciation is proportional to the capital stock. The average life of capital (AvgLifeOfCapt) varies across the technologies. Acquisitions of capital are proportional to the level of capital under construction (ElecCaptUndCons). Different construction delays are assumed depending upon the type of power plant technology. Capital stock under construction increases by the new investments made by the IPPs (IPPInvstElecCapt) and decreases when the construction of the power plants is completed. In the absence of demand for new capital, the construction rate reflects the compensation for the depreciated capital on an ongoing basis. The mathematical equations for the electricity capital sector are given in Box # 5.9.

[6] MW: Mega Watt.

Box # 5.9: Mathematic Equations for Electricity Capital Sector

```
ElecCapital = .0357|t=Coal; 9.9475|t=Hydro;1.22|t=Oil
;6.91|t=Gas
ElecCapital = +dt*ElecCaptAcq-dt*ElecCaptDepr
ElecCaptAcq = ElecCaptUndCons(t)/ConsDel
ElecCaptUndCons = ElecCaptDepr*ConsDel
ElecCaptUndCons = -dt*ElecCaptAcq+dt*ElecCaptCons
ElecCaptCons = ActualInvst/PriceOfElecCapt
ElecCaptDepr = ElecCapital/AvgLifeElecCapt
ConsDel = [4,6,3,4]
AvgLifeElecCapt = [30,40,25,30]
```

5.3.4 Resource Sector

The import dependency premium (ImportDependencyPremium) and the indigenous resource availability (IndigResourceAvailability) contribute to the major outputs of this sector.

The electricity demand and the safety margin (short-term coverage) for the on-site resource availability generate the desired demand for each type of resource (DesInv). Until indigenous depletion sets in, the desired resource inventory should be (is) met by the indigenous available resource stock (AvailResInv). The indigenous resource supply is determined by the ratio of indigenously available inventory (AvailResInv) to the desired inventory (DesInv).

The on-site inventory can be filled up to the resource storage capacity (ResStorageCap). The storage capacity is built based on resource inventory demand. The building of storage capacity takes place with a construction delay (CapBuidTime). When the indigenous resource is depleted, the on-site resource inventory is made up of the imported resource (Imports). These imports determine the import dependency (ImportDependency). However, the case of hydro resource (water) is different. The dam reservoir receives water from rain (RainFall), depending on the average duration of rainfall per year (AvgTimeForRain). Based on our desired dependency level (DesImportDependency), a premium (ImportDependencyPremium) is charged to gain a dependency reduction (DependencyToBeReduced). The premium, as shown in Fig. 5.7, is also one of the results of our study. We have identified this premium to imply that there is a preference for the least import-dependent technology. We consider this finding reasonable. Box # 5.9: The mathematic equations for Resource Sector are listed in Box # 5.10.

In this sector, the electricity production (ElecProdRate) for each of the electricity-generating technologies is determined, based upon the capacity utilization of the power plants (CapctyUtliz) and the total electricity production capacity of each of

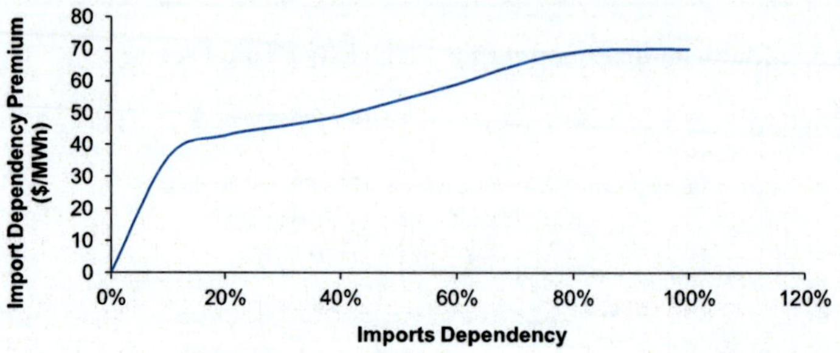

Fig. 5.7 Import dependency premium

the power plants (ElecProdCapacity). In our model, the production of electricity is driven by demand, which exerts its influence through the capacity utilization of the power plants. The electricity production capacity is constrained by the resource potential (ResPotProdElec) and the capital potential (CaptPotProdElec), which are the constituents of the total electricity production capacity. Both these factors are described in the following sub-sections. The equations for the electricity production sector are listed in Box # 5.11.

Box # 5.10: Mathematic Equations for Resource Sector

```
ImportDependencyPremium=GRAPHCURVE(DependencyToBeReduced,
0,0.1,[0,48.4,50.1,52.6,
54.6,55.5,56.3,56.6,55.2,56.6,58"Min:0;Max:60"])
DependencyToBeReduced=(ImportDependency-
(ImportDependency* (1-DesImportDependendcy)))
ImportDependency = IF(Imports > 0,ElecProdRate/
(ARRSUM(ElecProdRate)),0)
Imports=(0WHENt=Hydro;IF(InvGap>0,(IF(InvAdjstRate<=0,
(IF(InvGap<=CapLimit,
InvGap/InvAdjstTime,CapLimit/InvAdjstTime)),0)),0))
OnSiteResInv = [4178131,18,693,499,458,583,911,68101700]
OnSiteResInv  =  +dt*Imports  -dt*ResDepl+dt*RainFall  +
dt*InvAdjstRate
Imports=(0WHENt=Hydro;IF(InvGap>0,(IF(InvAdjstRate<=0,
(IF(InvGap
<=CapLimit,InvGap/InvAdjstTime,CapLimit/InvdjstTime)),0)),0))
InvGap = MAX(0,(DELAYINF(DesInv(t),1,3)-OnSiteResInv))
DesInv = ResDemand*Short_Term_Coverage
ResDemand= ((DemandOfElecFromEachTech
*ResConsumpRatio))/FuelEfficiency
InvAdjstRate=IF(Res_Remain>.2,(0WHENt=Hydro;IF
(AvailResInv>(TIMESTEP*InvGap)/InvAdjstTime,
(IF(InvGap<=CapLimit,InvGap/InvAdjstTime,CapLimit/
InvAdjstTime)) ,0)),0)
Res_Remain = AvailResInv/OrigIndigAvailRes
```

```
AvailResInv = OrigIndigAvailRes
AvailResInv = -dt*InvAdjstRate
InvAdjstRate=IF(Res_Remain>.2,(0WHENt=Hydro;IF
(AvailResInv>(TIMESTEP*InvGap)/InvAdjstTime,
(IF(InvGap<=CapLimit,InvGap/InvAdjstTime,
CapLimit/ InvAdjstTime)),0)),0)
IndigResourceAvailability = 1/(DesInv/AvailResInv)
CapLimit = MAX(0,(ResStorageCap-OnSiteResInv))
ResStorageCap = [4178131,18,693,499,458,583,911,68101700]
ResStorageCap = +dt*CapBuildRate
CapBuildRate          =          MAX((DELAYINF(DesInv(t),3,3)-
ResStorageCap)/CapBuidTime,0)
ResDepl = ElecProdRate*ResConsumpRatio
RainFall=IF(InvGap(Hydro)<=CapLimit(Hydro),InvGap(Hydro)
/AvgTimeForRain,CapLimit(Hydro))
OrigIndigAvailRes = [5000e2,5000e20,19,271,240,000,2e22]
InvAdjstTime = 1
CapBuidTime = [1,6,1,1]
```

Box # 5.11: Mathematical Equations for Electricity Production Sector

```
ElecProdRate = (ElecProdCapacity*CapUtiliz)
ElecProdCapacity = MIN(FuelPotToProdElec,CaptPotToProdElec)
CapUtiliz=IF(DemandOfElecFromEachTech(t)/Potential
ProdOfElec(t)>=1,1,IF(DemandOfElecFromEachTech(t)=
0,0,(DemandOfElecFromEachTech(t)/PotentialProdOfElec(t))))
```

5.3.5 Electricity Production Sector

5.3.5.1 Electricity Production Capacity

In our model, the electricity production capacity (ElecProdCapcity) of each of the readily available power plants is dependent upon two factors, the resource (fuel) potential (ResPotProdElec) and the electricity generating capital potential (Capt-PotProdElec) (e.g., utilization factor of the production capacity varies due to local operating conditions of electricity-generating power plants). The electricity production capacity is the minimum of the resource potential and the capital potential. The equation for the Electricity Production Capacity is listed in Box # 5.12.

Box # 5.12: Mathematical Equations for Electricity Production Capacity

```
ElecProdCapacity = MIN(FuelPotToProdElec,CaptPotToProdElec)
```

5.3.5.2 Resource Potential

The resource potential to produce electricity (ResPotProdElec) is determined based upon the onsite available stock of fuel (OnSiteResInv) to be burnt/exhausted, the fuel consumption ratio, the short-term inventory coverage, and the fuel efficiency. When electricity is generated, the onsite fuel stock is being depleted. The mathematical model for the Resource Potential is given in Box # 5.13.

Box # 5.13: Mathematical Equations for Resource Potential

```
FuelPotToProdElec=(OnSiteResInv/(ResConsumpRatio*
Short_Term_Coverage))/ FuelEfficiency
Short_Term_Coverage = .5
FuelConsumpRatio = [1.079,2,81.001,12.029]
FuelEfficiency = .4
```

5.3.5.3 Capital Potential

The electricity-generating potential (CaptPotProdElec) of each of the technologies is dependent upon the available electricity generating capital (ElecCapt) and the productivity of this capital (ElecCaptProductivity). In our model, the production potential is proportional to the electricity-generating capital as the unit capital productivity is assumed to be fixed for each technology. Consequently, we assume there is no technology (efficiency) improvement of significance. The electricity-generating capital (ElecCapt) is described in the (electricity) capital sector of the model. The equation for the Capital Potential is given in Box # 5.14.

Box # 5.14: Mathematical Equations for Capital Potential

```
CaptPotToProdElec = ElecCapital*ElecCaptProdctivity
```

5.3.5.4 Capacity Utilization

The capacity utilization of each of the power plants (CapctyUtliz) is a function of the ratio between the demand for electricity (DemandElecFromTech) and the capacity to produce electricity (TotElecProdCapcty). If this demand/supply ratio exceeds unity, then the capacity utilization function allows the power plant to run at 100% capacity. But when this ratio drops below unity, the capacity utilization is reduced in proportion to demand. The mathematical equations for the Capacity Utilization are listed in Box # 5.15. It should be noted that nuclear power plants have the highest capacity factor among all the electricity-generation technologies (i.e., the capacity factor of 93.3%). The highest capacity of a nuclear power plant primarily is due to low maintenance and refueling shutdowns (ONE, 2021)—another competing factor favoring the addition of nuclear power in the electricity supply mix of a country.

Box # 5.15: Mathematical Equations for Capacity Utilization

```
CapUtiliz=IF(DemandOfElecFromEachTech(t)/ElecProdCapacity
(t)>=1,1,IF(DemandOfElecFromEachTech(t)=0,0,
(DemandOfElecFromEachTech(t)/ElecProdCapacity(t))))
```

5.3.6 *Environment Sector*

The environment premium (EnvPrem) is the main output of this sector. Typically, the thermal generation carries a positive premium while the hydro generation carries a negative premium. The thermal premium (ThermalPrem) is computed based on the relative unit CO_2 cleanup cost (RelativeUnitCost) for each of the technologies, as described in Fig. 5.8.

The hydroelectricity premium (HydroElecPrem) is based on the average thermal premium to reflect the compensation for the competing hydro-based generation. CO_2 unit clean-up cost (UnitCleanupCosts) is instantaneously determined, based upon the clean-up operating costs (CleanupOperatingCosts) incurred on account of electricity cleaning (CleanElecRate). This cleanup process is described by a co-flow structure[7] presented in Fig. 5.9.

Electricity generation (thermal) causes CO_2 emissions. Depending on the volume of the emissions to be mitigated, a tax is charged (TaxIncomeRate). We have assumed that the more CO_2 a technology emits, the heavier it will be taxed relative to other

[7] In a co-flow structure, both the stocks (e.g., production capital and coressoponding energy intensity) more together in a proportion (Qudrat-Ullah & Davidsen, 2001; Sterman, 2000).

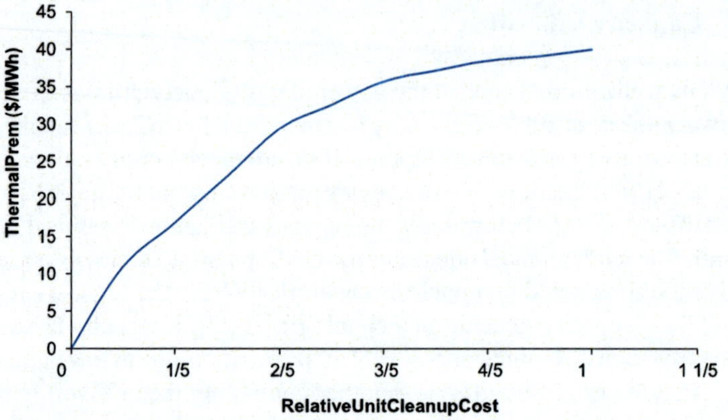

Fig. 5.8 Thermal premium

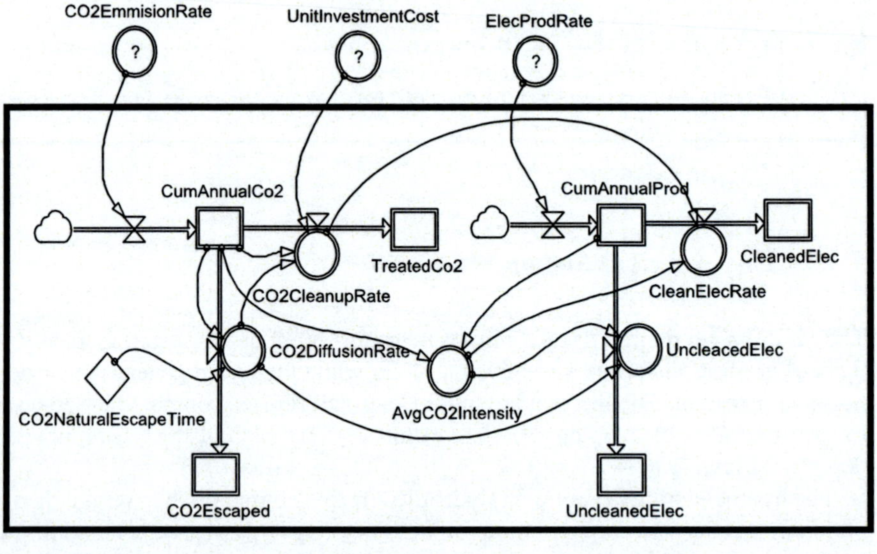

Fig. 5.9 Co-flow structure of clean electricity

technologies. This reflects the fact that mitigation costs for a higher level of CO_2 emissions are higher as well. A margin for mitigation investments (TrendedCleaupOperatingCosts) is also included in the tax rate. The CO_2 tax income (CumTaxIncome) is invested, over some time, to mitigate the emissions.

However, not all of the emitted CO_2 is treated/contained (sequestered). Some amount of emitted CO_2 escapes (CO_2 Escaped) beyond the national boundaries. This escaped CO_2 is a major concern in international mitigation negotiations. The co-flow structure helps us keep track of the amount of electricity in both the streams: one for which mitigation investments are being made and the other containing escaped CO_2.

No investments are made for the escaped CO_2. In addition, the average electricity-related CO_2 intensity (AvgCO2Intensity) is calculated through this structure. The larger the thermal electricity generation, the more CO_2 is emitted into the environment. The mathematical equations for the Environment Sector are listed in Box # 5.16.

Box # 5.16: Mathematical Equations for Environment Sector

```
EnvPrem = (IF(TIME<=1980,0,(HydroElecPrem |t =
Hydro;ThermalPrem)))
HydroElecPrem = -ARRAVG(ThermalPrem)
ThermalPrem=(GRAPHCURVE(,0,0.1,[0,10.9,16.8,23,29.4,32.6
,35.7,37.2,38.5,39.1,40"Min:0;Max:40"]))
RelativeUnitCost=IF(ARRSUM(UnitCleanupCosts)<=0,0,
UnitCleanupCosts/ARRSUM(UnitCleanupCosts))
UnitCleanupCosts=(IF(CleanElecRate>0,(0WHEN
t=Hydro;CleanupOperatingCosts/CleanElecRate),0))
CleanElecRate= 0|t=Hydro;CO2CleanupRate/AvgCO2Intensity
CO2CleanupRate=0|t=Hydro;(IF(CumAnnualCO2>((Cleanup
OperatingCosts)/(UnitInvestmentCost)-
CO2DiffusionRate),((CleanupOperatingCosts)/(UnitInvestment
Cost)),0))CumAnnualCO2 = CO2EmmisionFrac*[122,2,3,4]
CumAnnualCO2 = -dt*CO2DiffusionRate+dt*CO2EmmisionRate
-dt*CO2CleanupRate
CO2DiffusionRate=0|t=Hydro;IF(CumAnnualCO2>
(CumAnnualCO2/NaturalEscapeTime),
(CumAnnualCO2/NaturalEscapeTime),0)
CO2EmmisionRate = CO2EmmisionFrac*ElecProdRate
CleanupOperatingCosts=AvailInvestForCO2Cleanup*
EffectOfInvestAvailibilityOnCleanupCosts
AvailInvestForCO2Cleanup = CumTaxIncome/TimeToSpendTaxes
CumTaxIncome = DesTaxIncome
CumTaxIncome = -dt*CleanupOperatingCosts + dt*TaxIncomeRate
TaxIncomeRate = IF(DesTaxIncomeRate>0,DesTaxIncomeRate,0)
DesTaxIncomeRate=((TrendedCleaupOperatingCosts*
CleanupOperatingCosts)+CleanupOperatingCosts)+ TaxIn-
comeDiscrepancy
TrendedCleaupOperatingCosts = TREND(CleanupOperatingCosts,
5,1)
TaxIncomeDiscrepancy=(IF((DesTaxIncome-CumTaxIncome)>0,
(DesTaxIncome-CumTaxIncome)/TimeToAdjustTaxIncome,0))
EffectOfInvestAvailibilityOnCleanUpCosts=IF
(InvestRateforCO2Cleanup>0,
(IF(AvailInvestForCO2Cleanup>=InvestRateforCO2Cleanup,
InvestRateforCO2Cleanup/AvailInvestForCO2Cleanup,1)),0)
InvestRateforCO2Cleanup = UnitInvestment-
Cost*CO2ToBeCleaned)/TimeToReduceCO2
UnitInvestmentCost = IF(CO2ToBeCleaned>0,
(CO2MitigationCostCurve*10^9)/ CO2ToBeCleaned,0)
CO2MitigationCostCurve=GRAPHCURVE
```

```
(CO2ToBeCleaned/(3.667*10^9),0,5,
[0,14,26,45,69,90,110,140,160,190,220,250,290"Min:0;Max:300"
])CO2ToBeCleaned=IF((CumAnnualCO2+(CO2EmmisionRate*TIMESTEP))
CO2TargetLimit>0,CumAnnualCO2+(CO2EmmisionRate*TIMESTEP)-
CO2TargetLimit,0)
AvgCO2Intensity=0|t=Hydro;(CumAnnualCO2)/(ComAnnualProd);
CumAnnualProd =[31273.20,8,714,010,183,120,6053160];
CumAnnualProd = -dt*UncleanedElec -
dt*CleanElecRate+dt*ElecProdRate;UncleacedElec=0|t=Hydro;
CO2DiffusionRate/AvgCO2Intensity;
TimeToReduceCO2   =   3;   CO2TargetLimit   =   20.20;
CO2NaturalEscapeTime = 6; TimeToSpendTaxes = 3; TimeToAd-
justTaxIncome = 3
```

5.3.7 *Costs and Pricing Sector*

This sector keeps track of prices and costs of electricity. It should be noted that the price of electricity is often modeled as an expectation function. That is, the price of electricity (PriceOfElec) is the smoothed value of the required price (DesPriceOf-Elec), over an adjustment period (TimeToAdjsPriceOfElec) of 2 years. The required price is the price required to cover the unit cost of electricity production (Unit-CostOfElecProd) with a certain gross margin (GrossMrgn). The unit cost of electricity production consists of the sum of (i) fuel cost (FuelCost) and (ii) the operating costs (OpertCost). The gross margin is assumed to cover all of the capital costs (including the fixed operating and maintenance costs), the target rate of return on capital investments (TargetRetnOnInvest), and the increased marginal cost due to the capacity constraints (IncreasedMargnlCost). The rate of return on investments is assumed to be equal to the competitive industry/ interest rate, which in Pakistan's case is 8%. The fuel cost is the reference resource price (RefResourcePrice), modified by the market conditions. The resource supply and demand conditions are incorporated into the multiplier from the effect of the market on the price of the resource (EffectOfResAvailibilityOnPrice). This multiplier compares demand with supplies (IndigResourceAvailability). When demand exceeds supply, the multiplier is greater than one and the price is above reference resource price. The reverse is also true; oversupply causes depressed prices. The price of electricity is used by the economy to form a decision about its future consumption and investment in the electricity supply sector.

The net present value of investments (NPVOfInvst) is also determined in this sector. It depends on the net income (NetIncome) of the electricity generating technology, the average physical life of the capital (AvgLifeElecCapt), and the interest rate (IndustryRate) associated with competitive investments. The technology with longer capital life has more NPV of investment. The net income is the difference between the Govt. purchase price of electricity (GovPurchasePriceOfElec) and the required price of electricity (PriceOfElec), determined endogenously based on the cost of electricity production. In this formulation, two policy incentives are modeled: (i) how much is the purchase price offered by the Govt., (ii) what is the duration for which the Govt. guarantees the purchase of electricity. The relative NPV of investment in a technology determines the attractiveness of the investment incentives. In this mechanism, the best price-competitive electricity-generating technology is awarded the highest premium. A higher inventment incentive premium can lead to a higher market share of the respective technology. The mathematical equations for the Costs and Pricing Sector are given in Box # 5.17.

Box # 5.17: Mathematical Equations for Costs and Pricing Sector

```
NPVOfInvst=((((1+IndustryRate)^AvgLifeElecCapt)-1)/ (((1+
IndustryRate)^AvgLifeElecCapt)*IndustryRate))*NetIncome
NetIncome = ((GovPurchasePriceOfElec*10)-
(PriceOfElec))*ElecProdRate
PriceOfElec = (DELAYINF(DesPriceOfElec(t),
TimeToAdjsPriceOfElec))
DesPriceOfElec = (GrossMrgn+(UnitCostOfElecProd))
UnitCostOfElecProd = (FuelCost)+(OpertCost*10)+
UnitCleanupCleanupCost)*switch1
FuelCost = ResPrice; ResPrice = RefResour-
cePrice*EffectOfResAvailibilityOnPrice
EffectOfResAvailibilityOnPrice=EXP(ResourcePriceElasticity
*LN(IndigResource Availability))
GrossMrgn=(ElecCapital*(PriceOfElecCapt*(TargetRetnOnInvest
+(1/AvgLifeElecCapt))+ FOM_Cost))/ElecProdRate
GrossMrgn=((ElecCapital*(PriceOfElecCapt*(TargetRetnOn
Invest+(1/AvgLifeElecCapt))+FOM_Cost))/
ARRSUM(ActualElecProd))+IncreasedMargnlCost
IncreasedMargnlCost=(EffectofHydroDevelopmnt*(ElecCapital
*PriceOfElecCapt)/ (AvgLifeElecCapt))/ActualElecProd
EffectofHydroDevelopmn=GRAPHCURVE(ElecCapital(Hydro),0,15,
[0,0.001,0.014,0.02,0.035,0.041,0.051,0.061,0.075,
0.099,0.144,0.3"Min:0;Max:0.3"])|t=Hydro;0
GovPurchasePriceOfElec = 6.5;
IndustryRate = .08
TimeToAdjsPriceOfElec = 2;
OpertCost = [0.57,0.22,0.57,0.37]
RefResourcePrice = [1.01,0,1.97,0.23]
ResourcePriceElasticity = -.01
PriceOfElecCapt = [1.537,2,0.16700,0.518]
TargetRetnOnInvest = .06
FOM_Cost = [0.029,0.000005,0.0047,0037]
```

5.4 The Major Revisions and Assumptions of MDESRAP

As MDESRAP was originally developed in 2001 to assess Pakistan's energy policies (Qudrat-Ullah & Davidsen, 2001), it was natural to revise it and incorporate new features capable of accounting for the development regarding Pakistan's energy policies and their assessment regarding the commitments of the Government of Pakistan (as agreed in their INDC in 2016) about CO_2 reduction (INDC, 2016). The other objective, "to identify the best electricity supply mix, focused on renewable and nuclear power generation, for Pakistan necessitated the modifications to account for the additional policy incentives of the Energy Policy of Pakistan".

5.4.1 *The Key Modifications of MDESRAP*

Here are the major modifications of MDESRAP that are carried out to represent the current state of energy policies of Pakistan (Qudrat-Ullah, 2022):

i. To reflect the variability of substitution among completing generation technologies, the substitution mechanism is modeled as follows:

> *Depending upon the total cost (TC) of each technology, we apply a multinomial logit (MNL) model to obtain the share of each technology (ST). We focus, in addition to the standard cost elements, on the investment incentive premium (IP), the import dependency premium (DP), the environment premium (EP) for each technology, and we lump the rest in a power plant technology choice premium (TP). The standard cost element includes capital costs (CC), Operation and Maintenance costs (OC), and fuel costs (FC). We have built this modified total cost structure, with the inclusion of the three kinds of premiums, based upon the work of* (Moxnes, 1990)*]. The model structure allows the user to switch on or off any cost elements, as and when needed. Here is the mathematical form of it:*
>
> $$ST_i = \exp(-\alpha * TC_i) / \sum_i \exp(-\alpha * TC_i)$$
>
> *where,* $TC_i = CC_i + OC_i + FC_i + IP_i + DP_i + EP_i + TP_{i\,;i=1\ \mathrm{to}\ 7;}$ α is a distribution parameter and accounts for the local distribution of power plants.
>
> *When α takes an extreme value, then the aggregate choice of the whole sector (e.g., all the IPPs) for a technology mimics the choice of an individual IPP. The lower values of α represent greater variation between the individual choices and the aggregate (i.e., sector) choice.*

ii. *Instead of only three electricity-generating technologies (i.e., hydro, thermal, nuclear), the revised MDESRAP explicitly accounts for the dynamics of seven technologies: hydro, coal, gas, oil, nuclear, solar, and wind;*

iii. *Pakistan's Paris Agreement-based targets for the reduction of power-related* CO_2 *emissions are explicitly modeled;*

iv. *A life cycle approach for the assessment of power-related CO_2 emissions is modeled;*
v. *CPEC[8]-based power plants, predominantly coal-fired (e.g., Thar coal-based power plants), for the period 2020–26 are included;*
vi *Policy incentives both financial and non-financial are explicitly modeled (e.g., attractive bulk power tariffs and one-window approval process of new renewable producers)* (Qudrat-Ullah, 2022).

5.4.2 MDESRAP's Base Year's Parametric Values and Major Assumptions

As is the case of most modeling-based studies, a significant amount of data and its pretreatment was undertaken during the development of this revised MDESRAP. As RE2006[9] was the first major renewable policy in Pakistan that was implemented in the year 2008, therefore we selected the year 2008 as the base year for this study (Irfan et al., 2020). Table 5.1 lists Pakistan's power plant generation and other parametric values for each technology. Table 5.2 describes the cost data of the generation technologies. Key modeling assumptions are (Qudrat-Ullah, 2022):

Table 5.1 Power plants data (Akbar et al., 2017; IAEA, 2020; NEPRA, 2021; PEY13, 2014)

Power plant technology	Average life (years)	Generation (GWh)	Capacity (MW)[a]	Capacity factor (%)[a]	Utilization factor (%)[a]	CO_2 emissions (g CO_2/kWh)
Hydro	80	28,707	6480	98	46	12.8[a]
Oil[b]	30	30,818	8027	63	58	1502.86[b]
Coal[c]	30	136	50	80	23	790[c]
Gas[d]	30	32,923	40,970	85	60	531[d]
Nuclear	40	3077	462	79	85	12.51
Wind	20	0	0	33	22	11.38
Solar	25	0	0	35	29	11.38

[a]Capacity factor and utilization factor are average values based on the data of all the plants in operation in Pakistan (for Wind and Solar the source is Nicholas & Buckley, 2018); the majority of hydro generation is reservoir-based
[b]Majority of oil is furnace oil (imported) and large transportation distances are the factor in the higher emissions
[c]Majority of coal plants use locally available coal (transportation need is little)
[d]Majority of gas-based plants are single-cycle plants and use natural gas. Currently, nuclear waste is kept in pools at the site of the power plants in Pakistan, a relatively cheaper and low CO_2 emitting operation (WNA, 2020)

[8] CPEC: Cihina Pakistan Economic Corridoor.

[9] RE2006: It represent the Renewable Energy Ploicy of Pakistan that was promulgated in the year 2006.

Table 5.2 Cost of generation technologies (EIA, 2020; IRENA, 2019; Pachauri et al., 2014; Pey19, 2020)

Power plant technology	Capital Cost[a] ($/kW)	Fixed O&M cost ($/kW-year)	Variable O&M cost ($/MWh)	Key parametric values and assumptions
Hydro[b]	680	19.8	0.0	• RE and nuclear technologies' data is from IRENA, as they do not report on fossils • The Discount factor is 10% for the duration of this study • Data on Coal, Oil, and Gas power plants are based on IEA, EIA, PEY, and the least values are used to account for the local economic conditions (e.g., labor cost, transportation is cheaper)
Coal	1072	40.41	2.43[f]	
Gas[c]	536	54.4	14.04	
Oil	710	6.97	4.48	
Nuclear	2805	30	0.51	
Solar	618	39.56	0.0	
Wind[d]	1039	26.22	0.0	

[a]These are levalised costs (Jahangir et al., 2021)
[b]Reservoir-based hydropower data is used
[c]Majority of gas-based plants in Pakistan are single cycle
[d]Pakistan have predominantly onshore wind power plants; US$ are converted to Pakistani Rupees using an annual exchange rate as is reported by State Bank of Pakistan and vice versa (e.g., US$ = 59.86 in the year 2005–2006: Most of the coal used is local and costs less due to shorter distances and cheap labor (Fatima, & Nasim, 2019; Qudrat-Ullah, 2022)

- *T & D losses are assumed to decrease from 17.37% in 2013 to 12.5% in 2020, and about 11.7% by 2035, as projected by the NTDC (*2018*).*
- *Given the conditions of the power sector of Pakistan, the reserve margin is assumed as 10% for the modeling period 2080–2035.*
- *Two nuclear power plants each with 1100 MW capacity were to be operational in 2020 and 2021 but are delayed now and can be available by 2025.*
- *An additional coal-based generation capacity of 6600 MW by 2025, as per NTDC's projection regarding CPEC projects, is considered (*NTDC, 2018*).*
- *The ARE policy of 2019's target of 20% RE generation in the supply mix by 2025 is also accounted for in the model.*
- *GDP of Pakistan is taken as an external variable and the World bank database is used.*
- *Cost data for generation technologies are based on levalized costs.*
- *As per NTDC's projections, a carbon tax of US$ 31/tCO_2 is considered effective from July 1st, 2025 (*NTDC, 2018*).*

- *The annual exchange rate as published by the State Bank of Pakistan is used to cover US$ into Pakistan Rupees and (*vice versa*) and US$ is indexed to the base year.*

5.5 Summary of Chapter 5: The Key Insights

To understand the dynamics of energy policy in the context of ongoing challenges of climate change and energy security, to overcome several methodological limitations and constraints of the traditional modeling approaches, and provide an integrated modeling approach capable of representation of the dynamics of energy, economy, and the environment, system dynamics modeling approach is argued to be a viable solution. The development of a system dynamics model, MDESRAP together with its mathematical equations and corresponding explanation of the sector-wise (i.e., Electricity Demand sector, Investment sector, Capital sector, Resource sector, Production sector, Environment sector, and Costs and Pricing sector) structural elements of MDESRAP with the inter-sectoral interactions is described. In contrast to the optimization models, here in MDESRAP, a disequilibrium modeling framework, various sectorial dynamics are modeled at the level of diverse profit-seeking investor decisions (i.e., IPPs) incentivized by the policy under bounded rationality. In such a non-equilibrium approach, the policy intervention influences the economic development trajectory that policymakers need to monitor and evaluate on an ongoing basis (EC, 2016). This integrated modeling framework enables us to explore the outcomes of particular energy policy incentives and initiatives for generation technologies diffusion, electricity generation, CO2 emissions, and affordability and reliability of electricity as a macroeconomic change. The connection of diffusion and substitution of various generation technologies to a nonequilibrium model of the national energy system opens a very rich world of multidimensional micro and macro-level dynamics where impacts of energy policy reveal complex interactions between the energy sector, environment, and the economy.

The key modifications, assumptions, and parametric specifications to accurately capture the state of current energy policies and initiatives of Pakistan are explicitly listed in this chapter. Key insights are:

- System dynamics modeling methodology, due to its ability to capture and fully account for the dynamics of energy, economy, and the environment, stands out from the competing traditional methodologies (e.g., econometric models and input–output models).
- Modeling community in general and system dynamics modelers, in particular, can benefit from the various models of the key structures of any energy system. The use of exiting and validated sub-structures of a model, like our model (i.e., MDESRAP) adds to the structural validity of the model.
- Although our dynamic model, MDESRAP, is calibrated with the data of Pakistan's case, its modular structure is general enough to allow it to model any other electricity supply system of a country or region.

References

Akbar, M., Thaheem, M., & Arshad, H. (2017). Life cycle sustainability assessment of electricity generation in Pakistan: Policy regime for a sustainable energy mix. *Energy Policy, 111*, 111–126.

Al-Sarihi, A., & Bello, A. (2020). Socio-economic and environmental implications of renewable energy integrity in Oman: Scenario modeling using system dynamics approach. In H. Qudrat-Ullah & A. Kayal (Eds.), *Climate change and energy dynamics in the Middle East*. Springer.

Altaf, M. (2018). Renewable energy; identifying potential and development of solar energy to contribute to sustainable energy delivery in Pakistan. *International Journal of Engineering and Advanced Technology, 7*(3), 28–33.

Barlas, Y. (1989). Multiple tests for validations of system dynamics type of simulation models. *European Journal of Operational Research, 42*(1), 59–87.

Bassi, M., Deenapanray, P., & Davidsen, P. (2013). Energy policy planning for climate-resilient low-carbon development. In H. Qudrat-Ullah (Ed.), *Energy policy modeling in the 21st century*. Springer.

Dyner, I. (2000). Energy modelling platforms for policy and strategy support. *The Journal of the Operational Research Society, 51*(2), 136–144.

European Commission (EC). (2016). *Policy-induced energy technological innovation and finance for low-carbon economic growth.* Deliverable D2 Study on the Macroeconomics of Energy and Climate Policies. https://ec.europa.eu/energy/sites/ener/files/documents/ENER%20Macro-Energy_Innovation_D2%20Final%20%28Ares%20registered%29.pdf

EIA. (2020). The U.S. Energy Information Administration. *Levelized cost and levelized avoided cost of new generation resources in the annual energy outlook 2020.* https://www.eia.gov/outlooks/aeo/pdf/electricity_generation.pdf. Accessed 11 Jan 2021.

Fatima, U., & Nasim A. (2019). *Cost of electricity generation in Pakistan—Comparison of levelized cost of electricity of CPEC coal plants with oil and natural gas based plants commissioned in 2010–14* (Working Paper No. 02–19). https://ideaspak.org/wp-content/files_mf/1569051386IDEAS_WP2_Final.pdf. Accessed 10 Jan 2021.

Forrester, J.W. (1961). *Industrial dynamics.* MIT Press.

Gholizad, A., Ahmadi, L., Hassannayebi, E., Memarpour, M., & Shakibayifar, M. (2017). A system dynamics model for the analysis of the deregulation in electricity market. *International Journal of System Dynamics Applications, 6*(2), 1–30.

IAEA. (2020). *Country nuclear power profiles 2020 edition.* https://www-pub.iaea.orgMTCD/publications/PDF/cnpp2020/countryprofiles/Pakistan/Pakistan.htm. Accessed 12 Jan 2021.

INDC. (2016). *Pakistan's intended nationally determined contribution.* Ministry of Climate Change, Government of Pakistan. http://mocc.gov.pk/PublicationDetail/ZGM3YjM3MDEtMjcxOS00MDNjLTliMTMtNTk2MmNhZDRiMWMw. Accessed 12 Dec 2021.

International Renewable Energy Agency (IRENA). (2019). *Renewable Power Generation Costs in 2019.* IRENA, Abu Dhabi. https://www.irena.org/-/media/Files/IRENA/Agency/Publication/2020/Jun/IRENA_Power_Generation_Costs_2019.pdf. Accessed 11 Jan 2021.

Irfan, M., et al. (2020). Assessing the energy dynamics of Pakistan: Prospects of biomass energy. *Energy Reports, 6*, 80–93.

Jahangir, M., Mostafaeipour, A., Habib, H., Saghaei, H., & Waqar, A. (2021). Effect of emission penalty and annual interest rate on cogeneration of electricity, heat, and hydrogen in Karachi: 3E assessment and sensitivity analysis. *Journal of Engineering, 6679358*, 1–16. https://doi.org/10.1155/2021/6679358

Moxnes, E. (1990). Interfuel substitution in OECD-European electricity production. *System Dynamics Review, 6*(1), 44–65.

Muller, O., Kaufmann-Hayoz, R., Schwaninger, M., & Ulli-Beer, S. (2013). The diffusion of eco-technologies: A model-based theory. In H. Qudrat-Ullah (Ed.), *Energy policy modeling in the 21*st *century*. Springer.

NEPRA. (2021). *Generation licenses granted to power plants by NEPRA*. Generation licenses granted to power plants by NEPRA. https://nepra.org.pk/. Accessed 26 Jan 2021.

Nicholas, W., & Buckley, T. (2018). The Institute for Energy Economics and Financial Analysis. Pakistan's Power Future. http://ieefa.org/wp-content/uploads/2018/11/Pakistans-Power-Future_December-2018.pdf. Accessed 20 Nov 2021.

NTDC. (2018). *Electricity demand forecast based on multiple regression analysis*. National Transmission and Distribution Company, Government of Pakistan. http://www.mocc.gov.pk/SiteImage/Policy/National%20Climate%20Change%20Policy%20of%20Pakistan%20(2).pdf. Accessed 13 Nov 2021.

Office of Nuclear Energy (ONE). (2021). *Nuclear power is the most reliable energy source and it's not even close*. https://www.energy.gov/ne/articles/nuclear-power-most-reliable-energy-source-and-its-not-even-close. Accessed 24 Aug 2021.

Pachauri, K., Allen, R., Barros, R., Broome, J., Crame, W., Christ, R., Church, A., Clarke, L., Dahe, Q., & Dasgupta, P. (2014). *Climate change 2014: Synthesis report*. Contribution of Working Groups I, II, and III to the Fifth Assessment Report of the Intergovernmental Panel on Climate Change; IPCC: Geneva, Switzerland.

PEY13. (2014). Pakistan Energy Yearbook. (2013). *Ministry of Petroleum and Natural Resources, Islamabad, Pakistan*.

PEY19. (2020). Pakistan Energy Yearbook. (2019). *Ministry of Petroleum and Natural Resources, Islamabad, Pakistan*.

Qudrat-Ullah, H. (2008). Behavior validity of a simulation model for sustainable development. *HomeInternational Journal of Management and Decision Making, 9*(2), 1290140.

Qudrat-Ullah, H. (2013). Energy policy modeling in the 21st century: An introduction. In H. Qudrat-Ullah (Ed.), *Energy policy modeling in the 21*st *century*. Springer.

Qudrat-Ullah, H. (2015). Independent power (or pollution) producers? Electricity reforms and IPPs in Pakistan. *Energy, 83*(1), 240–251.

Qudrat-Ullah, H. (2016). *The physics of stocks and flows of energy systems*. Springer.

Qudrat-Ullah, H. (2022). A review and analysis of renewable energy policies and CO_2 emissions of Pakistan. *Energy, 238*(Part B), 121849.

Qudat-Ullah, H., Akrofi, M., & Kayal, A. (2020). Analyzing actors' engagement in sustainable energy planning at the local level in Ghana: An empirical study. *Energies, 13*(8), 2028.

Qudrat-Ullah, H., & BaekSeo, S. (2010). How to do structural validity of a system dynamics type simulation model: The case of an energy policy model. *Energy Policy, 38*(5), 2216–2224.

Qudrat-Ullah, H., & Davidsen, P. (2001). Understanding the dynamics of electricity supply, resources, and pollution: Pakistan's case. *Energy, 26*(6), 595–606.

Rashid, I., Benhelal, E., & Rafiq, S. (2020). Reduction of greenhouse gas emissions from gas, oil, and coal power plants in Pakistan by carbon capture and storage (CCS): A review. *Chemical Engineering Review, 43*(11), 2140–2148.

Rehman, U., Cai, Y., Siyal, Z. A., Mirjat, N. H., Fazal, R., & Kashif, S. U. R. (2020). Cleaner and sustainable energy production in Pakistan: Lessons learnt from the Pak-times model. *Energies, 13*(108), 1–21.

Sterman, J. (2000). *Business dynamics: Systems thinking and modeling for a complex world*. McGraw-Hill.

Uddin, R., Shaikh, A. J., Khan, H. R., Shirazi, M. A., Rashid, A., & Qazi, S. A. (2021). Renewable energy perspectives of Pakistan and Turkey: Current analysis and policy recommendations. *Sustainability, 13*, 3349. https://doi.org/10.3390/su13063349

Valasai, D., Uqaili, A., Memon, H., Samoo, R., Mirjat, N., & Harijan, K. (2017). Overcoming electricity crisis in Pakistan: A review of sustainable electricity options. *Renewable and Sustainable Energy Reviews, 72*, 734–745.

Verwiebe, P. A., Seim, S., Burges, S., Schulz, L., & MüllerKirchenbauer, J. (2021). Modeling energy demand—A systematic literature review. *Energies, 14*, 7859. https://doi.org/10.3390/en14237859

World Nuclear Association (WNA). (2020). *Nuclear power in the world today*. https://www.world-nuclear.org/information-library/current-and-future-generation/nuclear-power-in-the-world-today.aspx. Accessed 15 Sep 2021.

Ziras, C., Calearo, L., & Marinelli, M. (2021). The effect of net metering methods on prosumer energy settlements. *Sustainable Energy, Grids and Networks, 27*, 100519.

Chapter 6
Calibration, Initialization, and Validation of the Simulation Model

6.1 Introduction

The degrees to which a model functions as intended and behaves similarly to the real-world target system are important considerations in most of the model-based decision making processes. Energy policymaking is essentially a complex dynamic task. To make better decisions in these complex systems, modeling and simulation have long been used. We need models to make most of the organizational decision making situations. Therefore, testing and validation of a model is a pre-requite for its use as a decisional aid.

Since its inception, the system dynamics community has been a strong advocate of the validation of models. for a system dynamics-based model such as MDESRAP, validation is 'the process of establishing confidence in the soundness and usefulness of a model' (Forrester & Senge, 1980). There are two major categories of validation: (i) structural validity—*a sound traceable structure is responsible for the behavior (outcome of the model*, (ii) behavioral validity—*model outcomes are comparable to the real system (that is being modeled)*. In the repertoire of system dynamics, researchers have developed various statistical and theoretical tests and procedures for the validation of their simulation models, Some common procedures for the structural validity of a simulation model are (i) theoretical structural validation, and (ii) structurally-oriented behavior tests (i.e., Extreme Conditions test, Behaviour Sensitivity test, and Modified-Behaviour Prediction test). For behavior validity, the system dynamics community have often relied on various statistical procedure including:

i. Trend comparison and removal
ii. Autocorrelation function test for period comparison
iii. Cross-correlation function test for phase lag detection
iv. Comparing the means
v. Comparing the amplitude variations

H. Qudrat-Ullah, *Understanding the Dynamics of Nuclear Power and the Reduction of CO_2 Emissions*, https://doi.org/10.1007/978-3-031-04341-3_6

The validity of our model, MDESRAP was ascertained by subjecting it to this battery of tests. The purpose of this chapter, therefore, (i) describe the initialization process and calibration process for MDESRAP, (ii) to demonstrate the utility of various validation tests by the way of applying them to MDESRAP.

6.2 Model Calibration and Initialization

Calibration is the process to find the best match between simulated and observed values. The model has been calibrated to represent the electricity generation/production history (from 2006 to 2021) of Pakistan. This period is considered, mainly, because the Government of Pakistan introduced its renewable-focussed energy policy, RE 2006 Policy, in 2006 (Qudrat-Ullah, 2015). Subsequently, three new policies: (i) 2012 NCCP[1] of Pakistan, (ii) National Power Policy of 2013, and (iii) ARE[2] Policy of 2019 was introduced to promote cleaner energy production. The government of Pakistan declared the special investment incentives for the renewable and cleaner electricity generation sector through these policies (referred to as "energy policy of Pakistan") that are to be analyzed in this study.

Model initialization is the process whereby its parameters are set their base values. The model is initialized to be in a steady-state under the influence of the exogenous economic conditions (GDP of Pakistan), with the base year 2006. This general condition imposes certain restrictions upon the choice of initial values for some of the state variables (stocks) and some of the parameters. Therefore, we apply analytic initialization. This initialization is described in Sect. 6.2.1. The remaining state variables and parameters are assigned explicit values to comply with the calibration of the model, described in Sects. 6.2.2 and 6.2.3, respectively.

6.2.1 Analytic Initialization[3]

6.2.1.1 Electricity Capital Under Construction (ElecCaptUndCons)

The electricity capital under construction, i.e., the capital that becomes ready to generate electricity only after the construction time over (i.e., construction is complete), is initialized concerning the capital being depreciated;

$$\text{init} \rightarrow \text{ElecCaptUndCons} = \text{ElecCaptDepr} \times \text{ConsDel} \tag{6.1}$$

[1] NCCP: National Climate Change Policy

[2] ARE: Alternate Renewable Energy

[3] The variables contained in smoothes and delays are initialized to the value that is endogenously computed in the model and is to be smoothed or delayed.

This initial condition is derived in the following way: According to the model, the capital depreciation represents the electricity-generating capital over the average physical lifetime of the capital;

$$\text{ElecCaptDepr} = \text{ElecCapital}/\text{AvgLifeElecCapt}; \tag{6.2}$$

Substituting (6.2) into (6.1), we obtain the initial value applied during the initialization.

6.2.1.2 Resource Storage Capacity (ResStorageCap)

The resource storage capacity is initialized, under steady-state condition, to be the desired resource inventory[4];

$$\text{init} \rightarrow \text{ResStorageCap} = \text{DesInv},$$

The electricity demand and the safety margin (short-term coverage) for the on-site resource supply determine the desired resource inventory (DesInv). By tracking and substituting all the variables and parameters involved, we obtain the initial value applied during the initialization.

6.2.1.3 On-Site Resouce Inventory (OnSiteResInv)

The on-site resource inventory, i.e., the fuel to be consumed during the electricity production, is initialized also to be the desired inventory;

$$\text{init} \rightarrow \text{OnSiteResInv} = \text{DesInv}$$

6.2.1.4 Indigenous Resource Inventory (AvailResInv)

The indigenous resource inventory is initialized to be the estimated original quantity of resources of Pakistan in the year 1980;

$$\text{init} \rightarrow \text{AvailResInv} = \text{OrigIndigAvailRes}$$

[4] init = Initialization.

6.2.1.5 Investment in Electricity Capital (ActualInvst)

The actual investment in electricity generating technology is initialized, under steady-state condition, to be the indicated investments;

$$\text{init} \rightarrow \text{ActualInvst} = \text{IndicInvstInElecCapt}$$

The indicated investments are determined by the total demand for production capital (TotDesElecCapt) and the share (ActualShareTech) of each of the competing technologies. By tracking and substituting all the variables and parameters involved, we obtain the initial value applied during the initialization process.

6.2.1.6 Energy Intensity (EnrgyIntensity)

The energy intensity of GDP of Pakistan is initialized to be the reference energy intensity for the base year (1980) and its value is 0.53 (tons of oil equivalent per thousand dollars of GDP);

$$\text{init} \rightarrow \text{EnrgyIntensity} = \text{RefEnrgyIntensity}$$

6.2.1.7 Average Electric Intensity (AvgElecIntensity)

The average electricity intensity is initialized to be the reference electricity intensity of the base year and is 0.487 (kWh per thousand dollars of GDP);

$$\text{init} \rightarrow \text{AvgElecIntensity} = \text{RefElecIntensity}$$

6.2.1.8 Capital Intensity (ElecIC)

The capital intensity is initialized, under the steady-state assumption, concerning the long term indicated electricity intensity and the electricity generating capital;

$$\text{init} \rightarrow \text{ElecIC} = \text{LongTermIndcElecIntensity} \times \text{ElecCapital}$$

The reference electricity intensity (RefElecIntensity) and the effect of electricity price on electricity intensity (EffectOfPriceOnElecIntensity) determine the long-term indicated electricity intensity. By tracking and substituting all the variables and parameters involved, we obtain the initial value applied during the initialization.

6.2.1.9 Capital Intensity-under-Construction (ElecICon)

The capital intensity-under-construction that determines the current electricity intensity (when the capital is added, the electricity intensity increases and upon its depreciation, intensity experiences a decrease), is initialized, under the steady-state assumption, concerning the long term indicated electricity intensity and the electricity generating capital-under construction;

$$\text{init} \rightarrow \text{ElecICon} = \text{ElecCaptUndCons} \times \text{LongTermIndcElecIntensity}$$

Here again, the reference electricity intensity (RefElecIntensity) and the effect of electricity price on electricity intensity (EffectOfPriceOnElecIntensity) determine the long-term indicated electricity intensity. By tracking and substituting all the variables and parameters involved, we obtain the initial value applied during the initialization.

6.2.1.10 Cumulative Tax Income

The cumulative tax income, under the steady-state assumption, is initialized to be the desired tax income;

$$\text{Init}\quad \text{CumTaxIncome} = \text{DesTaxIncome}$$

The desired tax income is dependent upon the investments being made for mitigation of emissions (InvestRateforCO2Cleanup) and the time to spend the tax income (TimeToSpendTaxes). By tracking and substituting all the variables and parameters involved, we obtain the initial value applied during the initialization.

6.2.2 Value Initialization

In this section, we present a summary of the values assigned explicitly to the state variables remaining after the analytic initialization (Table 6.1).

6.2.3 Parameter Values

In this section, we give a summary of the values assigned to the parameters of the model remaining after analytic initialization (Table 6.2). The main data comes from the case unless mentioned otherwise.

Among these parameters, the values of the following four parameters are derived by a specific procedure as explained below;

Table 6.1 State variables with assigned values

Variable	Name	Value	Comments
Electricity capital	ElecCapital	[0.047,13.12,0.27,9.19] (MW)	These values are assigned to match the capital's potential for electricity generation with the historic electricity production of the base year (1980)
Share of each electricity-generating technology	ShareOfTech	[0.0021,0.58,0.012,0.406] (fraction)	These values are assigned to match the historic electricity production by each of the technology for the base year (1980)
Cumulative electricity production	CumAnnualProd	[31273.20,8,714010, 183120,6053160] (MWh)	These values are taken to match with the amount of electricity produced in the base year (1980)
Cumulative CO_2	CumAnual CO_2	[21247,0,202,811, 4159171] (tons [CO_2])	These values reflect the base year emissions in addition to some previous accumulation
Escaped CO_2	CO_2Escaped	0 (Tons [CO_2])	No escaped CO_2 (assumed for simplicity)
Uncleaned electricity	UncleanedElec	0 (MWh)	–
Cleaned electricity	CleanedElec	0 (MWh)	–
Mitigated CO_2	TreatedCo_2	0 (Tons [CO_2])	–
Cumulative cleanup operating cost	CumCleanupOperatingCost	0 ($)	Nothing is spent for mitigating the CO_2 emissions

[a]The values given in such an array will always refer to Coal, Hydro, Oil, and Gas respectively

i. Effect of Electricity Price on Its Demand

We derive the value of this parameter from the model. The environment sector and the resource sector are decoupled. Then, the model is exposed to the reference electricity intensity of GDP, and the effect of electricity price on its demand is chosen to obtain the best fit between simulated and historic electricity intensity.

Table 6.2 Parameters with assigned values

Variable	Name	Value
Time to Adjust Investments	AdjTimeForInvst	2 (years)
Average Physical Life of Capital	AvgLifeElecCapt	[30,40,25,30] (years)
Average Time to Rain	AvgTimeForRain	0.32 (years)
Time to Build Resource Storage Capacity	CapBuidTime	[1,6,1,1] (years)
CO_2 Emission Fraction	CO2EmmisionFrac	[0.953,0,0.807,0.733] (tons [CO_2])/MWh)
Time for CO_2 to Escape the Boundary	CO2NaturalEscapeTime	6 (years)
Target Limit for CO_2 Emissions	Co2TargetLimit	20.20 Million tons (CO_2)
Construction Delay for Power Plant	ConsDel	[4,6,3,4] (years)
Capacity Factor for Power Plant	CpacityFactor	[0.85,9,8,86] (fraction)
Desired Resource Import Dependency	DesImportDependendcy	0 (fraction)
Distribution Parameter (α)	DistPara	0.02488 (MWh/$)
Effect of Electricity Price on its Demand	EffectOfElecPriceOnDemand	−1.56 (fraction)
Effect of Energy Price on its Demand	EffectEnrgyPriceDemand	−0.04 (fraction)
Electricity Capital Productivity	ElecCaptProdctivity	3,504,000 (MWh/(MW-Year)
Fuel Consumption Ratio	FuelConsumpRatio	[1.079,2.000,81.001,12.029] (Btu/MWh)
Fuel Efficiency	FuelEfficiency	0.4 (%)
Fixed Operating and Maintenance Cost	FOM_Cost[a]	[0.029,0.000005,0.0047,0.0037] ($/MWy)
Gov.'s Purchase Price of Electricity	GovPurchasePriceOfElec	6.5 (0.01$/kWh)

(continued)

Table 6.2 (continued)

Variable	Name	Value
Interest Rate in Competitive Industry	IndustryRate	0.08 (%)
Inventory Adjustment Time	InvAdjstTime	1 (year)
Operating Cost	OpertCost	[0.57,0.22,0.57,0.37] ($/MWh)
Original Indigenous Resources (Ready to be transported at consumption sight)	OrigIndigAvailRes[b]	[[6.06428e15,2e25,4,563,440,1.0584e15] (Btu)
Premium for Employability, Flexibility, Availability of the Resource	OtherPrem	[170, –44, 97, −26] ($/MWh)
Price of Electricity Capital	PriceOfElecCapt[c]	[1.537,2,0.167,0.518] ($/MW)
Reference Electricity Intensity	RefElecIntensity	0.50 (kWh/$)
Reference Energy Intensity	RefEnrgyIntensity	0.53 (Toe/000$)
Reference Resource Price	RefResourcePrice	[1.01,0,1.97,0.23] (0.01$/kWh)
Effect of Resource Availability of its Price	ResourcePriceElasticity	– 0.01 (fraction)
Safety Margin for Resource Inventory	Short_Term_Coverage	0.5 (year)
Target Rate of Return on Investments	TargetRetnOnInvest	0.08 (%)
Time to Adjust Energy Intensity	TimeAdjEnrgyIntensity	2 (years)
Time to Adjust Electricity Price	TimeToAdjsPriceOfElec	2 (years)
Time to Adjust Electricity Capital	TimeToAdjstElecCapt	6 (years)
Time to Adjust Tax Income	TimeToAdjustTaxIncome	3 (years)

(continued)

Table 6.2 (continued)

Variable	Name	Value
Time to Establish (Reference) Electricity Price for Investments in Oil	TimeToEstRefElecPrice	3 (years)
Time To Observe the Sustained Change in Electricity Price	TimeToObserveSustainedChng	2 (years)
Time to Realize Change in Electricity Intensity	TimeToRealizeChng	3 (years)
Time to Reduce CO_2 Emissions to Target	TimeToReduceCo2	3 (years)
Time to Spend Tax Income	TimeToSpendTaxes	3 (years)

[a]Derived from Schrattenholzer and Schafer (1996)
[b]Taken from EIDADB (1994)
[c]The capital cost for hydro-based capacity is taken from and for the rest of the technologies, the capital cost comes from Schrattenholzer and Schafer (1996)

ii. **Effect of Energy Price on Its Demand**

This parameter is also estimated from the model. After decoupling the environment sector and the resource sector, the model is exposed to the reference energy intensity of GDP, and the effect of energy price on its demand is chosen to obtain the best fit between simulated and historic energy intensity.

iii. **Resource Price Elasticity**

The resource price elasticity is also derived through model simulation. The model, after decoupling the environment sector, is exposed to the reference resource price. The resource price elasticity is chosen to obtain the best fit between simulated fuel price and historic import price of fuel.

iv. **Distribution Parameter (α)**

The coefficient for the distribution of investment α is set according to rough estimates of the variation in the cost elements of electricity generation technology, in Pakistan. Because of a significant variation in the fuel costs at various power plant locations, the fuel costs are chosen as indicative of all cost elements. We considered all 19 locations of thermal power plants, where the fuel is consumed to generate the electricity. The fuel costs at each of these sites were obtained and based on fuel cost differences (from their average value); we obtained 7.97 (c/kWh) as the average variation. This implies a 26.57 (c/kWh) variation in total costs because of generation efficiencies of about 30 percent. In other words, these cost differences extend 13.28 (c/kWh) to each side of the mean costs. We use this estimate as our starting point when we assume the standard deviation of the Weibull distribution to be 26.57. This estimate yields a value of $\boldsymbol{\alpha} = \mathbf{0.2488}$ **(kWh/c[5])** using the formula $\boldsymbol{\alpha^2 = 6/\sigma^2}$.

v. **Other Premium**

This parameter value is selected to match the model-generated value of the share of investments with the base year value.

6.3 Model Validation

Validation is the process of building confidence in the usefulness of a model (Forrester & Senge, 1980; Qudrat-Ullah, 2017). To establish such validity, a system dynamics model must generate the "right output behavior for the right reasons" (Barlas, 1996). *In system dynamics models, the model structure can be compared to descriptive knowledge of real-system structure; and model behavior may be compared to observed real-system behavior* (Forrester & Senge, 1980; Qudrat-Ullah, 2008). In the following Sects. 6.3.1 and 6.3.2, our model is exposed to some tests for its structural as well as behavioral validity.

6.3.1 Structural Validity of the Model

Since SD models are causal models (Barlas, 1989; Qudrat-Ullah & BaekSeo, 2010; Sterman, 2000), the structural validity of a system dynamics model is of fundamental importance in the overall validation process. We have applied a two-prong approach towards the structural validation of our model. Firstly, during the construction of the model, we utilized (i) the specific case, Pakistan's data, and (ii) the sub-models/structures of the existing models of the domain. The relationships developed in the model, which were based on the specific case data, provided a sort of 'empirical' structural validation (Forrester & Senge, 1980). The adopted sub-models of the existing models of the domain served as a 'theoretical' structural validation (Forrester & Senge, 1980). Secondly, we applied the structurally oriented behavior tests to validate the structure of the model.

[5] c stands for cent = .01 US$

Table 6.3 Model structures adopted from the existing work

Sub-structure/Concepts	Reference	Remarks
Investment incentives dynamics	Dyner (2000)	The causal structure was adopted
Substitution mechanism between the electricity and the competitor (oil)	Davidsen (1986)	The structural formulation was adopted
Substitution mechanism among the competing electricity generating technologies	Moxnes (1990)	The structural formulation was adopted
Production capital structure	Moxnes (1990)	The structural formulation was adopted
Gross margin	Sterman (1984)	The structural formulation was adopted

6.3.1.1 Theoretical Structural Validation

In Table 6.3, we provide the sources of information used during the model's sub-structures development.

6.3.1.2 Structurally Oriented Behavior Tests

The function of structural validity tests is to check whether the structure of the model is an adequate representation of the real system (Barlas, 1989). Barlas has suggested three structurally oriented behavior tests: (i) Extreme Conditions test, (ii) Behaviour Sensitivity test, and (iii) Modified-Behaviour Prediction test. By examining the model behavior under different conditions, these tests help us detect the major structural flaws in a model (Barlas, 1989).

We apply (i) the Extreme Conditions test, (ii) the Behaviour Sensitivity test to examine the structural validity of our model. We do not apply the third test i.e., Modified-Behaviour Prediction test, due to lack of data regarding the behavior of some "modified" versions of the real system (which could be due to parameter modifications, or structural modifications). In these tests, our focus is on behavior patterns rather than numerical behavior validity.

Extreme Conditions test: In this test, extreme values are assigned to selected parameters and then the model-generated behavior is compared to the reference (or anticipated) behavior of the real system, under the same extreme condition. In our model, as an extreme condition, we set ConsDel (construction delay) to a very large number (like 10,000,000), which in reality means "no construction" of the new production capacity. The resulting behavior of the model is presented in Fig. 6.1.

The production capacity exhibits a steady decline pattern, eventually approaching the "zero level". The reason is obvious: the loop responsible for the addition of new production capacity breaks out. Under the influence of better economic conditions (increased GDP), there is always a call for investments in the production capacity; the stock of capital-under-construction continues to increase; almost zero out-flow

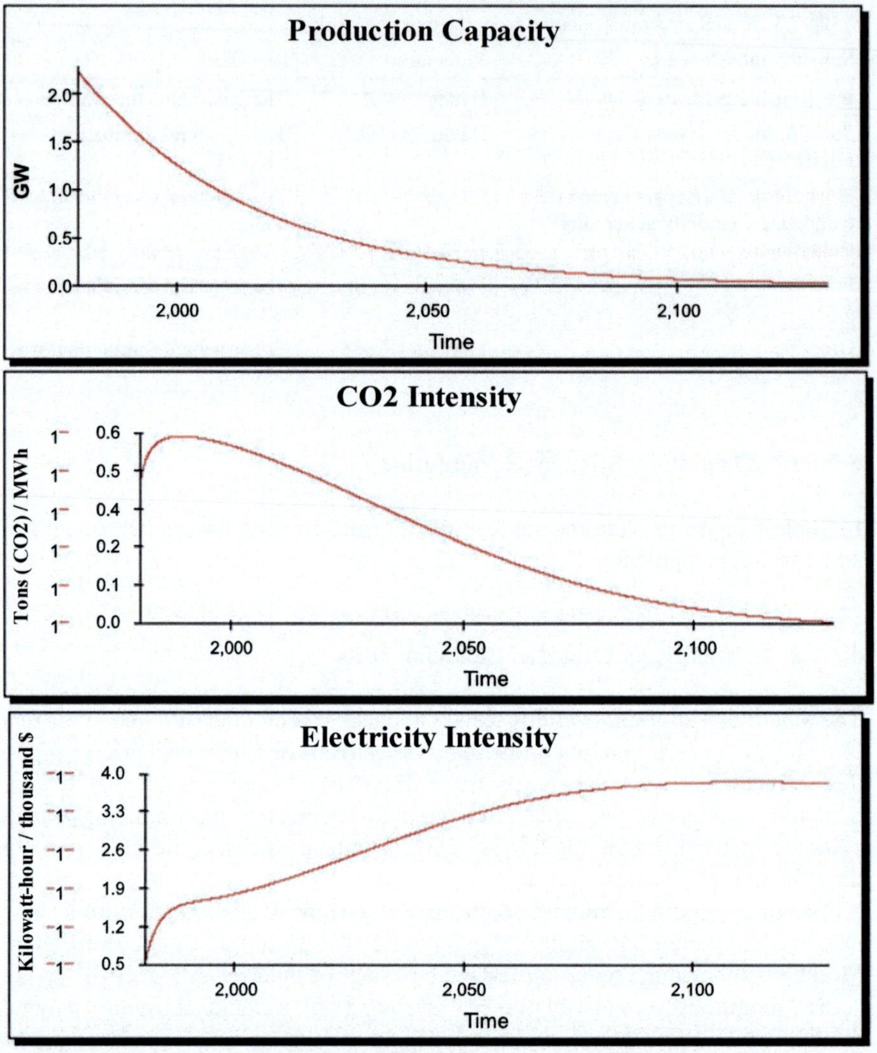

Fig. 6.1 Most behavior under extreme condition test

from the stock of capital-under-construction pushes this stock to an even higher level. But this higher level of capital-under-construction, in turn, makes the eventual total depletion (due to depreciation) of the existing capital (production capacity) stock follow a much slower trajectory.

The CO_2 intensity, after a little initial rise, follows the path, similar to that of the production capacity. The initial rise may be due to the substitution of the less capital-intensive technology i.e., oil-based power plants (they consume relatively high CO_2-intensive fuel). Eventually, when there is no production capacity, there is no electricity generation; no electricity-related CO_2 emissions. However, the electricity intensity (of GDP) follows a rather stabilized (at a value of 1.6 kilowatt-hour/thousand $) trajectory. This behavior pattern of the model is justified in light of the assumption that there is always a minimum part of energy consumption dependent upon electricity (though the price of the alternative is much cheaper). We observe that the model, under extreme conditions, exhibits a behavior that is in line with the anticipated behavior of the real system. Therefore, we conclude that the model passes the extreme condition test and its validity is enhanced.

Behavior Sensitivity test: In this test, the behavioral sensitivity of the model is evaluated, i.e. how does the model's behavior change over time as alternative parameter values are tested in the model, one at a time? When we applied the test, the fundamental patterns of behavior (of production capacity, electricity intensity, and CO_2 intensity) proved to be rather insensitive to most of the system parameters. The parameters were tested with both an increase and a decrease of 50% (to its base value). The results of the sensitivity tests are listed in Appendix 1. Changing these parameters affected only the specific numerical values of behavior characteristics (such as an earlier take-off, or shortened doubling time). Such numerical sensitivity is in line with the anticipated behavior of the real system under the same conditions. As a result, the more is our confidence in the model structure.

6.3.2 *Behaviour Validity of the Model*

Validation of model behavior is an important part of simulation validation in general, and System Dynamics model validation in particular (Barlas, 1989; Qudrat-Ullah & BaekSeo, 2010). The main purpose of behavior validation is to compare the model-generated behavior to the observed behavior of the real system. The behavior pattern of the electricity supply is assumed indicative of the model behavior. Barlas (1989) has suggested a set of pattern-oriented tests appropriate for the SD model's behavior validation:

i. Trend comparison and removal
ii. Autocorrelation function test for period comparison
iii. Cross-correlation function test for phase lag detection
iv. Comparing the means
v. Comparing the amplitude variations

MDERAP has been successfully subjected to these tests for its validity (Qudrat-Ullah, 2008; Qudrat-Ullah & BaekSeo, 2010), as is described below:

6.3.2.1 Trend Comparison and Removal

For this test, we applied the Growth Curve Model (exponential) to estimate the trend of both the model-generated and the reference data series. This model provided the relative best fit of the data. The plot of the trends is presented in Fig. 6.2. The

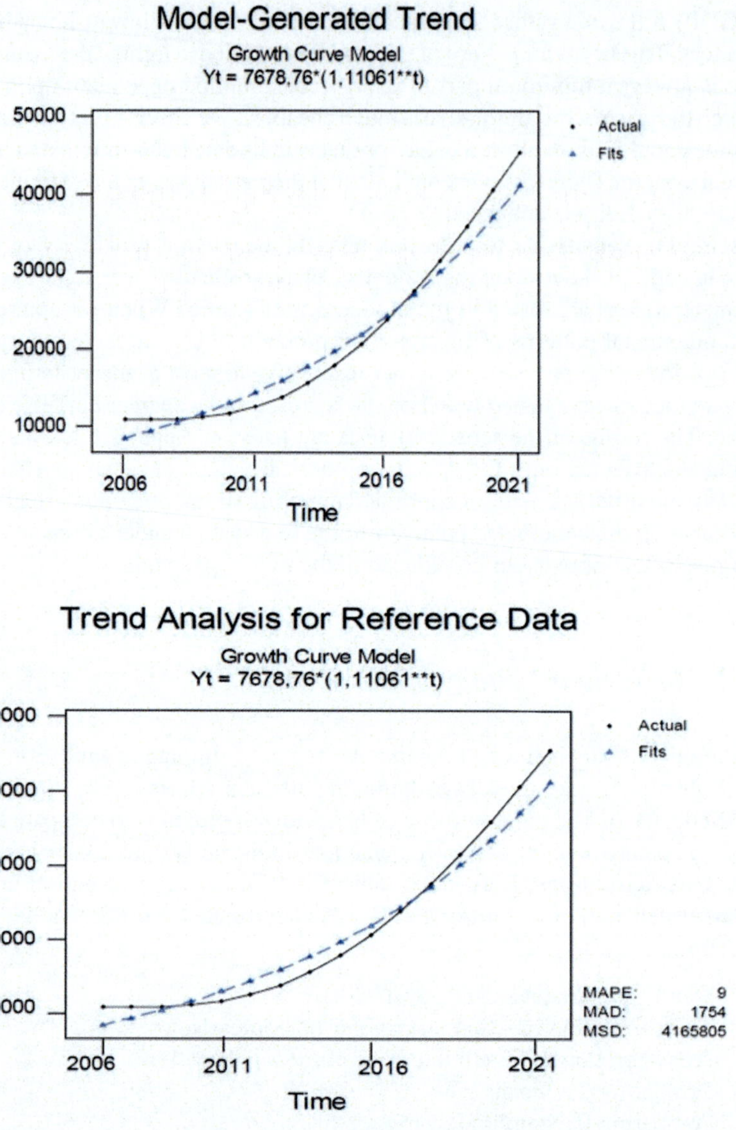

Fig. 6.2 Plot of trends of both the model-generated and the reference data

estimated trend coefficients for both the time patterns are given in Table 6.4. We see that the slope estimates for the reference and model-generated patterns are 1.10136 and 1.11061 respectively, which are not significantly different. Thus, the model passes this test. For the next steps of the pattern validation procedure, we de-trend both the time patterns by applying the transformation $Z_t = Y_t - Y_t$^ (Barlas, 1989).

Table 6.4 Trend equations and the estimated trend coefficients

Data	Trend equation	Trend coefficients (b_0, b_1)
Model-generated data	Y_t^ = 7678,76*(1,11061**t)	**7678.76 and 1.11061**
Reference data	Y_t^ = 8907,58*(1,10136**t)	**8907,58 and 1,10136**

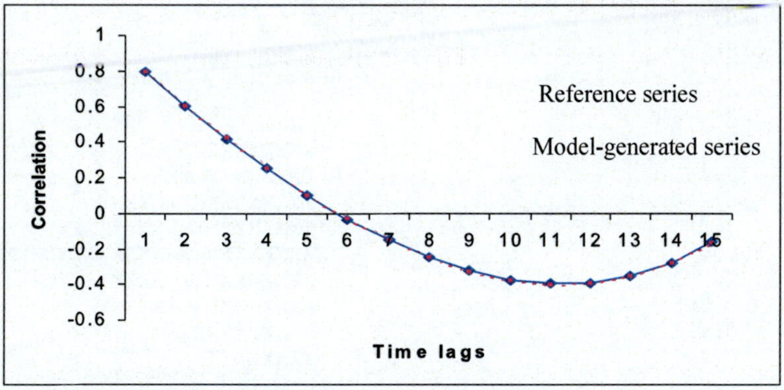

Fig. 6.3 Plot of autocorrelated functions of both (the detrended) series

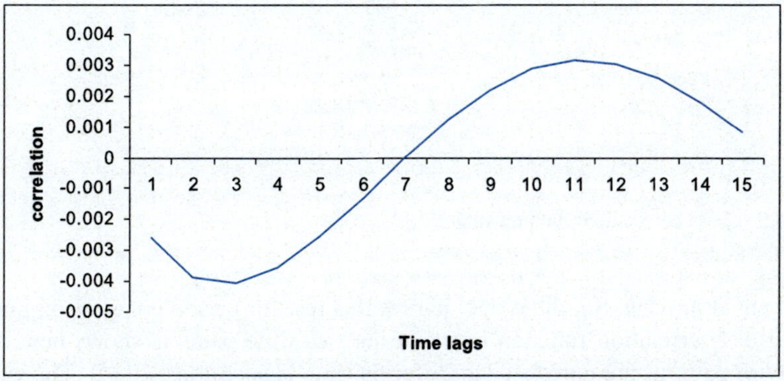

Fig. 6.4 Plot of the difference of autocorrelated functions

6.3.2.2 Autocorrelation Function Test for Period Comparison

The autocorrelation[6] test can detect significant errors in the periods (Barlas, 1989). In Fig. 5.3, we observe that neither of the time patterns (model-generated and reference) possesses high-frequency components (that affect only the very low lag autocorrelations).

Also in Fig. 6.4, we see that the autocorrelation functions are not significantly

[6] These tests may be more useful and relevant when the model behaviour is an oscillation.

```
           -1,0 -0,8 -0,6 -0,4 -0,2  0,0  0,2  0,4  0,6  0,8  1,0
             +----+----+----+----+----+----+----+----+----+----+
-15  -0,163                      XXXXX
-14  -0,280                   XXXXXXXX
-13  -0,355                 XXXXXXXXXX
-12  -0,393                XXXXXXXXXXX
-11  -0,397                XXXXXXXXXXX
-10  -0,371                 XXXXXXXXXX
 -9  -0,319                  XXXXXXXXX
 -8  -0,241                    XXXXXXX
 -7  -0,142                      XXXXX
 -6  -0,024                         XX
 -5   0,113                          XXXX
 -4   0,265                          XXXXXXXX
 -3   0,431                          XXXXXXXXXXXX
 -2   0,610                          XXXXXXXXXXXXXXXX
 -1   0,800                          XXXXXXXXXXXXXXXXXXXXX
  0   1,000                         XXXXXXXXXXXXXXXXXXXXXXXXXX
  1   0,793                          XXXXXXXXXXXXXXXXXXXXX
  2   0,598                          XXXXXXXXXXXXXXXX
  3   0,416                          XXXXXXXXXXX
  4   0,249                          XXXXXXX
  5   0,097                          XXX
  6  -0,037                         XX
  7  -0,153                      XXXXX
  8  -0,249                    XXXXXXX
  9  -0,322                  XXXXXXXXX
 10  -0,371                 XXXXXXXXXX
 11  -0,393                XXXXXXXXXXX
 12  -0,386                XXXXXXXXXXX
 13  -0,346                 XXXXXXXXXX
 14  -0,271                   XXXXXXXX
 15  -0,157                      XXXXX
```

Fig. 6.5 Cross-correlation function values

different at any lag. So, the model passes this test for period comparison purposes. The cross-correlation function between the two-time patterns shows how the two behavior patterns are correlated at different time lags (Barlas, 1989). The values of the cross-correlation function, which correlates the model-generated time pattern and the reference time pattern, are portrayed in Fig. 6.5. We see (in Fig. 6.5) that the cross-correlation function reaches a maximum value of 1.0 at **lag 0**: meaning no phase lag between the two behavior patterns. Thus, the model passes the cross-correlation function test for phase lag detection.

6.3.2.3 Comparing the Means

In general, both the simulated and the actual time patterns are highly correlated therefore, the standard *t-test* is not applicable for the comparison of means (Barlas, 1989). As a practical alternative, Barlas (Barlas, 1989) has suggested examining the

'Percent error in the means (E_1)' to see the discrepancy between the means. The means of the simulated and the reference time series are 20,999.93 and 22,301.0 respectively. The percent error in the mean (E_1) turns out to be 5%. Since the percent error in the means, 'E_1' is small, it seems that the model does not have a significant systematic error and hence, passes this test.

6.3.2.4 Comparing the Amplitude Variations

To compare the variations in the simulated output, Barlas has suggested the 'Percent error in the Variations (E_2)' as:

$E_2 = |S_S - S_A| / S_A$, where S_S and S_A are the standard deviations of the simulated and reference time patterns respectively. The values of the standard deviations of the simulated and the reference time series are 21,347.2858 and 21,347.2858 respectively. The percent error in the variations comes out to be 3%. This small percent error in the variations, 'E_2' implies that the model revision/modifications are not called for and the behavior pattern validity of the model is further reinforced.

6.3.3 *Theil's Inequality Statistics*

The mean-square error and Theil's inequality statistics have been used to evaluate the historical fit of our model. These statistics convey a measure of overall behavior discrepancy. The model is a dynamic general disequilibrium representation of Pakistan's electricity supply sector. Starting the model simulation in 2006 provides sixteen years of simulated data to compare to the actual behavior of the electricity supply sector of Pakistan. As indicated by Figs. 6.6, 6.7 and 6.8, the results of the simulation reproduce Pakistan's experience, concerning (i) energy intensity, (ii)

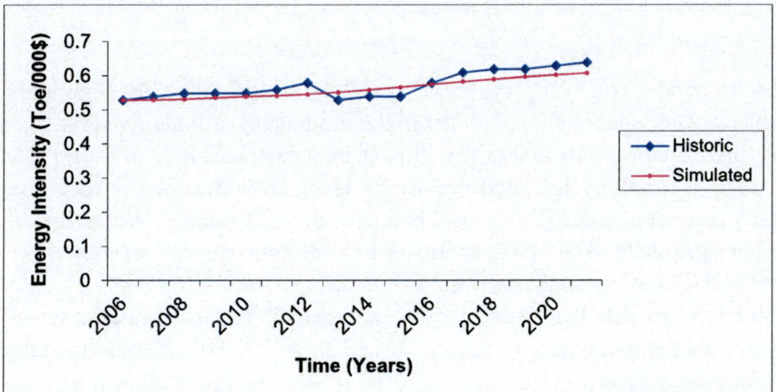

Fig. 6.6 Simulated and historic energy intensity

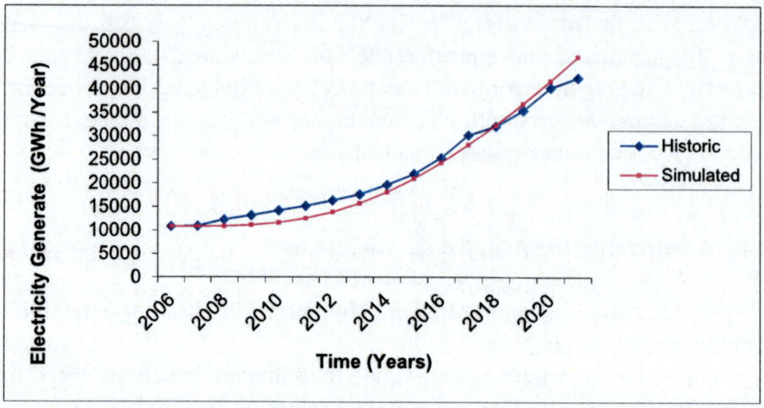

Fig. 6.7 Simulated and historic electricity generation

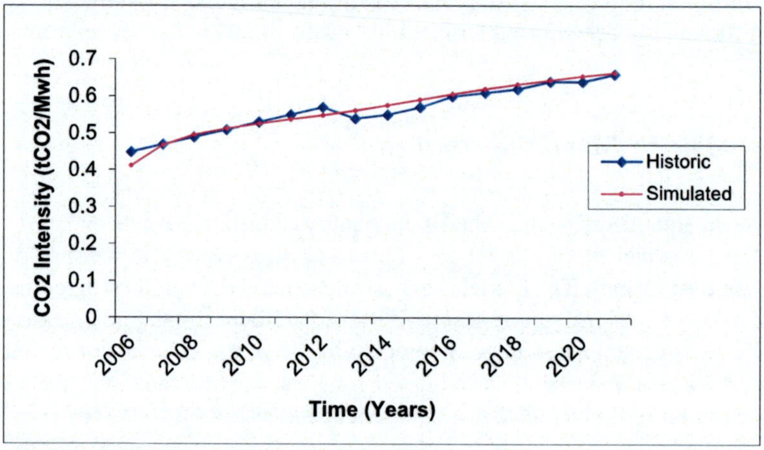

Fig. 6.8 Simulated and historic CO_2 intensity

electricity production (supply), and (iii) CO_2intensity, relatively accurately. These variables are endogenously generated in the model and sufficiently serve the purpose of our investigation: "to assess the impact of investment incentives on electricity-generating technology mix and emissions level, over the long-term". The energy intensity provides us with the overall behavior of the economy. While, the other two, electricity production and CO_2 intensity exhibit the scenarios of electricity supply and related CO_2 emissions in the environment.

Table 6.5 presents the error analysis in terms of the mean squared error (MSE), the root mean squared percent error (RMSPE), and Theil's inequality statistics for these three variables.

The RMS percent error provides a normalized measure of the magnitude of the error. The MSE and inequality statistics provide a measure of the total error and how

Table 6.5 Error analysis of the model

Variable	MSE (units^2)	RMSPE (%)	U^M	U^S	U^C
Energy intensity	0.00040144	4	0.08	0.17	0.75
Electricity production	3,842,542.62	10	0.13	0.54	0.35
CO_2 intensity	0.00026138	3	0.10	0.46	0.44

it breaks down into bias, unequal variation, and unequal co-variations components (Sterman, 1984). While the small total errors in the variables provide confidence in the model, large errors might suggest the presence of the internal inconsistency of the model or the particular structure controlling the variables (with large errors).

Theil's inequality statistics provide us with an overall error decomposition to resolve such doubts. As shown in Table 6.5, we have slected three variables of our model to perform error analysis. In the case of "energy intensity", the RMS percent error is 4%. This low error means that the model-geerated values of of this variable are similar ot the actucal data. The major portion (nearly 75%) of this 4% error is due to unequal co-variation. This indicates that the simulated energy intensity tracks the underlying trend in the historical energy intensity almost perfectly, but diverges point-by-point.

The 2nd variable, electricity production, has an RMS percent error of 10%. The error decomposition reveals that 13% of MSE is due to bias with 35% due to unequal co-variation and 54% due to unequal variation. The plot of error is presented in Fig. 6.9 and reveals that electricity production, after beginning with a shortfall of over 8% in 2006 did experience an improvement with a little rise of 2% in 2008. After 2009, the production again fell about 4% for the whole duration till 2011, where it again surpassed the actual production with a little scale (2–4%) for the rest of the period.

This behavior can be explained as due to the assumption about the size of power plants in our model. We have assumed the size of power plants to be 100 MWe.

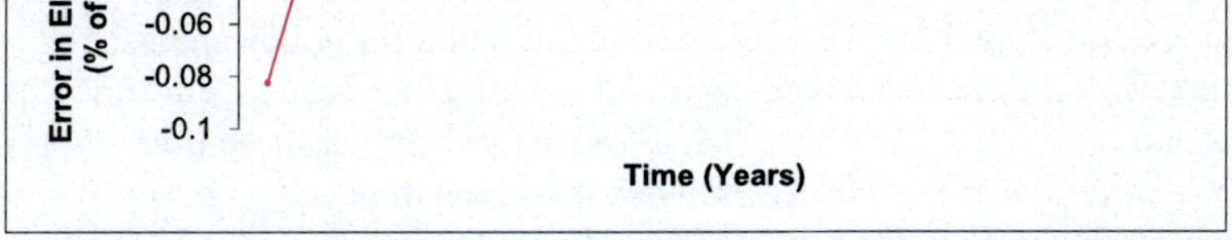

Fig. 6.9 Error in electricity production

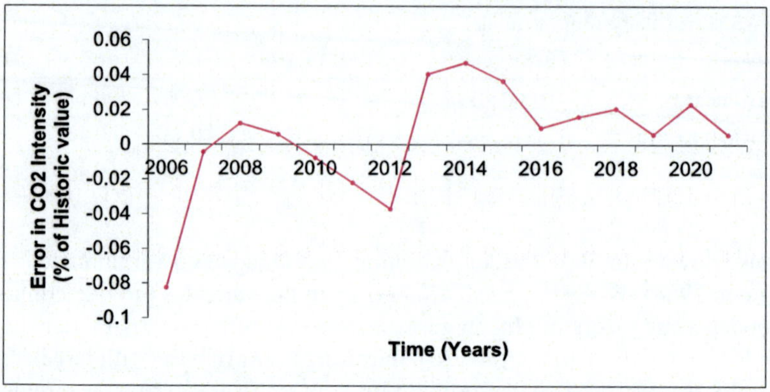

Fig. 6.10 Error in CO_2 intensity

However, the model could easily be "tuned" (by selecting the appropriate variable values for power plant size) to replicate the historic production of electricity. In the case of econometric and statistical models, which are often used for point prediction or short-term forecasting, such tuning would be appropriate and would help build confidence in the utility of the model. However, system dynamics models are primarily used for an assessment of long-term impacts and policy analysis, relying on such ad-hoc adjustments to the parameter to improve the replication of the historic data, would not enhance the validity of the model. Instead, there is a chance of decreasing confidence in the model by obscuring the model's ability to endogenously generate the behavior of interest (Qudrat-Ullah, 2022).

Finally, we consider the 3% RMS percent error in "CO_2 intensity." Here, only 10% of MSE is due to bias, and the rest is (almost) equally shared by unequal variance and unequal covariance. Again, a simplifying assumption appears to be responsible for this small error. The plot of error is presented in Fig. 6.10. We have assumed that the initial accumulation of CO_2 corresponds to the current electricity generation. But in reality, the accumulation of CO_2 stock could be higher due to prior cumulative accumulations (Qudrat-Ullah, 2022). In our model, it would be an easy matter to "tune" this value to generate the matching value for CO_2 intensity. Also, another assumption about the fuel intensity appears to be responsible for the over-estimation of CO_2 intensity generated by the model during the period 2012–2021. We have assumed a constant fuel intensity (consumption/ unit of electricity produced) throughout the whole period. However, by simple adjustments of the fuel intensity parameter, the model would replicate the actual CO_2 intensity. Since the model is not intended for point forecasting but rather for policy analysis, the error in CO_2 intensity is of little concern, as it will not affect the relative efficacy of policies.

For the behavioral validity of MDESRAP, we applied these *multiple tests* (*i.e., Trend comparison and removal, Autocorrelation function test for period comparison, Cross-correlation function test for phase lag detection, Comparing the means, and Comparing the amplitude variations and Theil's inequality statistics*). Each of these tests focuses on a specific pattern component of the generated outcome by a model. The model, MDESRAP, passed all the tests quite convincingly. Therefore, given that the model had already passed the structural validity phase of the model validation process, the favorable outcome of the multiple tests and Theils inequality statistics contribute significantly to enhancing the credibility of the model, MDESRAP. In the next chapter, we will use this validated model for our policy assessment and design objective.

6.4 Summary of Chapter 6: The Key Insights

The utility of any model, including MDESRAP, is heavily dependent on its credibility. The system dynamics community has developed a variety of tests and procedures to ascertain both the behavioral and structural validity of their simulation models. MDesrap is no expection. In this chapter, we have described the initialization and calibration process for MDESRAP and demonstrated the systematic validity of MDESRAP by subjecting it to is a battery of tests. For its structural validity three tests: (i) Extreme Conditions test, (ii) Behaviour Sensitivity test, and (iii) Modified-Behaviour Prediction test was applied and MDESRAP successfully met these standards for its structural validity.

Five tests: (i) Trend comparison and removal, (ii) Autocorrelation function test for period comparison, (iii) Cross-correlation function test for phase lag detection, (iv) Comparing the means, and (v) Comparing the amplitude variations together with Theils inequality statistics were applied to test the behavioral validity of MDESRAP. Given the positive outcome of these multiple tests, it is reasonable to consider MDESRAP as a credible model, fit to be used for policy analysis and design tasks. Modeling and simulation community in general and system dynamics and agent-based models can avail these tests to validate and increase the appeal and a wider acceptance of their models and model-based studies and resulting insights.

Appendix 1: Sensitivity Analysis for the Variable, "Electricity Demand"

Parameter	Base case value	Upward sensitivity value	Downward henergu value
AdjTimeForInvst	2 (years)	6206.4	6244.1
AvgLifeElecCapt	[30,40,25,30] (years)	6714.5	4892.6
AvgTimeForRain	0.32 (years)	6244.1	6244.1
CapBuidTime	[1,6,1,1] (years)	6244.1	6244.1
ConsDel	[4,6,3,4] (years)	6180.1	6121.1
EffectOfElecPriceOnDemand	−0.02 (fraction)	11,145.9	3322.5
ElecCaptProdctivity	3,504,000 (MWh/(MW−Year)	16,136.7	2207.3
FuelConsumpRatio	[1.079,2.000,81.001,12.029] (Btu/MWh)	6137.9	6242
FuelEfficiency	0.4 (%)	6048.7	6239.8
FOM_Cost	[0.029,0.000005,0.0047,0.0037] ($/MWy)	6199.6	6199.6
GovPurchasePriceOfElec	6.5 (0.01$/kWh)	6199.6	6199.6
IndustryRate	0.08 (%)	6199.6	6199.6
InvAdjstTime	1 (year)	6199.6	6199.4
OpertCost	[0.57,0.22,0.57,0.37] ($/MWh)	5584.1	6934.8
OrigIndigAvailRes	[5000e2,50e20,1,927,124,000,2e22] (Btu)	6201.6	6195.7
OtherPrem	[170, −44, 95, −26] ($/MWh)	6022	6461.5
PriceOfElecCapt	[1.537,2,0.167,0.518] ($/MW)	3743	13,092.6
RefElecIntensity	0.50 (kWh/$)	8141.7	5234.1
RefResourcePrice	[1.01,0,1.97,0.23] (0.01$/kWh)	5943.1	6486.9
ResourcePriceElasticity	−0.01 (fraction)	6229.8	6167
Short_Term_Coverage	0.5 (year)	6241.5	5981.2
TargetRetnOnInvest	0.08 (%)	4143.5	10,498.3

(continued)

(continued)

Parameter	Base case value	Upward sensitivity value	Downward henergu value
TimeToAdjsPriceOfElec	2 (years)	6235.5	6165.6
TimeToAdjstElecCapt	6 (years)	6172.2	6195.5
TimeToRealizeChng	3 (years)	6110.6	6280.3

References

Barlas, Y. (1989). Multiple tests for validations of system dynamics type of simulation models. *European Journal of Operational Research, 42*(1), 59–87.

Barlas, Y. (1996). Formal aspects of model validity and validation in system dynamics. *System Dynamics Review, 12*(3), 183–210.

Dyner, I. (2000). Energy modelling platforms for policy and strategy support. *The Journal of the Operational Research Society, 51*(2), 136–144. https://doi.org/10.2307/254253

Forrester, J. W., & Senge, P. M. (1980). Tests for building confidence in system dynamics models. *TIME Studies in the Management Science, 14*, 209–228.

Moxes E. (1990). Interfuel substitution in OECD–European electricity production. *System Dynamics Review, 6*(1), 44–65.

Qudrat-Ullah, H. (2008). Behavior validity of a simulation model for sustainable development. *International Journal of Management and Decision Making, 9*(2), 129–139.

Qudrat-Ullah, H. (2015). Independent power (or pollution) producers? Electricity Reforms and IPPs in Pakistan. *Energy, 83*(1), 240–251.

Qudrat-Ullah, H. (2017). How to enhance the future use of energy policy simulation models through ex-post validation. *Energy, 120*(1), 58–66.

Qudrat-Ullah, H. (2022). A review and analysis of renewable energy policies and CO_2 emissions of Pakistan. *Energy, 238*, 121849.

Qudrat-Ullah, H., & BaekSeo, S. (2010). How to do structural validity of a system dynamics type simulation model: The case of an energy policy model. *Energy Policy, 38*(5), 2216–2224.

Sterman, J. D. (1984, Winter). Appropriate summary statistics for evaluating the historical fit of system dynamics models. *Dynamica, 10*, 51–66.

Sterman, J. D. (2000). *Business dynamics: Systems thinking and modeling for a complex world*. McGraw-Hill.

Chapter 7
Development and Evaluation of CO_2 Reducing Scenarios

7.1 Introduction

Sustainable availability and use of energy play a fundamental role in our lives. However, in the case of the majority of the developing countries, there is a paradoxical situation: meeting the rising energy demand, using indigenous fossils, and reducing energy-related emissions at the same time. Pakistan, a developing nation, is faced with a similar difficult situation. To meet the rising electricity demand and bring in clearer and renewable electricity in the supply mix, the Government of Pakistan introduced various policies and incentives (the detail is provided in Chapter 1 of this book when we term it is, Energy Policy of Pakistan), we set the key objectives of this book as (i) whether or not the Energy Policy of Pakistan meets the pledged reduction of CO_2, and (ii) what is the best electricity supply mix for Pakistan to reduce electricity-related CO_2 emission and move toward a low-carbon economy. To address these questions, in this chapter we present (i) the development of CO_2 reducing scenarios, (ii) a systematic system-wide assessment of these developed scenarios in terms of reliable availability, affordability, and cleaner production of electricity, and (iii) identify and discuss the best possible pathways for the policymakers of Pakistan to meet the PA-based CO_2 reduction targets and move toward a low-carbon economic regime. We accomplish these objectives by utilizing our validated dynamic model, MDESRAP, in the scenario planning approach.

The remainder of this chapter is organized as follows: Sect. 7.2 describes the construction of CO_2 reducing scenarios, the policy assessment is carried out in Sects. 7.3, 7.4, and 7.5 presents the results of the evaluation of the alternate scenarios, through a comparative analysis of all the three scenarios, Sect. 7.6 showcase the best option for the policymakers and finally, Sect. 7.7 list the key insights for the readers and decision makers in the energy domain.

H. Qudrat-Ullah, *Understanding the Dynamics of Nuclear Power and the Reduction of CO_2 Emissions*, https://doi.org/10.1007/978-3-031-04341-3_7

7.2 CO_2 Reducing Scenarios Development

Scenario-based planning and policymaking is an authentic and scientific approach (Means et al., 2005). For instance, using a dynamic model, ENGAGE, generated socio-economic scenarios relevant to a given emissions target. Although pursuing a 100% renewable energy-based supply is easier said than done. For positive cases of a 100% renewable energy supply mix, please see scenario-based studies by Brown et al. (2018) and Hansen et al. (2019). For negative cases of a 100% renewable energy supply mix, please see the works of Trainer (2012) and Trainer (2013). Sadiqa et al. (2021), using scenario-based planning explored a fully renewable energy pathway for Pakistan's energy sector by 2050 (Sadiqa et al., 2021). In another study, using LEAP model-based scenarios, Mirjat et al. (2018) predicted the long-term electricity supply–demand for Pakistan (Mirjat et al., 2018). For a possible path toward a low-carbon economy in Colomia, Dyner et al. (2013) effectively demonstrated the utility of a system dynamics-based scenario planning approach (Dyner et al., 2013).

What is a scenario? According to the Intergovernmental Panel on Climate Change (IPCC) (IPCC, 2008):

> *A scenario is a coherent, internally consistent, and plausible description of a possible future state of the world. It is not a forecast; rather, each scenario is one alternative image of how the future can unfold.*

Based on this definition, we can use scenarios to provide a better understanding of a dynamic view of the future (Mahmoud et al., 2009). All policy decisions have both short-term and long-term consequences. When policymakers are striving to achieve the objective of having affordable, reliable, and cleaner electricity-based future for their people, a systematic scenario-based assessment of their energy policies can provide critical information and insights for the decision makers. Therefore, for the systematic, socio-economic and environmental, and system-wide assessment and design of the energy policy of Pakistan, we will use various scenarios including status-quo scenarios together with renewable and nuclear power-focused scenarios. General steps in this scenario-based process for energy policymaking are shown in Fig. 7.1.

As the policy objectives of the energy policy of Pakistan, the focus of this study, are described earlier in Chapter 2 of this book, here we begin the process of scenario-based energy policymaking, aimed at achieving affordable, reliable, and cleaner production of electricity in Pakistan, by constructing the following scenarios.

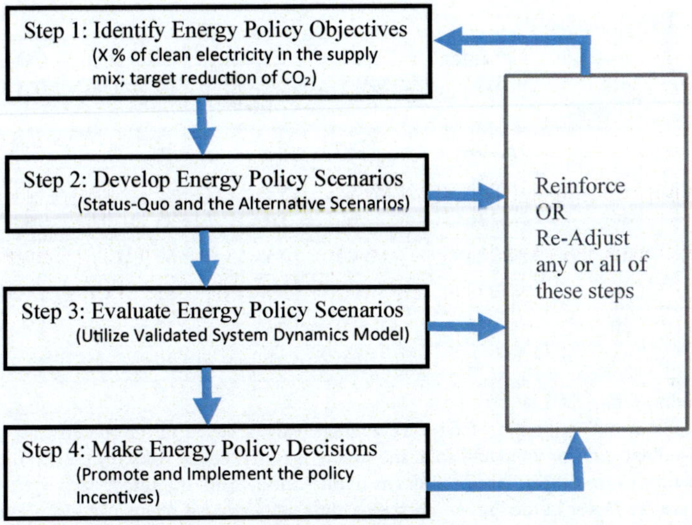

Fig. 7.1 Key steps in scenario-based energy policy decisions

7.2.1 *Status-Quo Scenario-Business as Usual (BAU) Scenario*

The first major renewable policy in Pakistan was promulgated in 2006 (i.e., RE 2006), therefore we selected the year 2008 (when this policy was implemented) as the base year for this study. Table 7.1 lists Pakistan's power plant generation and other parametric values for each technology. Table 7.2 described the cost data of the generation technologies. To account for the current energy policy and conditions of the energy sector in Pakistan, the key features of the status-quo scenario include (Qudrat-Ullah, 2022):

(i) *T & D*[1] *losses are assumed to decrease from 17.37% in 2013 to 12.5% in 2020, and about 11.7% by 2035* (NTDC,[2] 2018).

(ii) *The reserve margin is assumed as 10% for the modeling period 2080–2035.*

(iii) *Two nuclear power plants each with 1100 MW capacity, which was to be operational in 2020 and 2021 can be available by 2025* (PNR, 2018).

(iv) *An additional coal-based generation capacity of 6600 MW by 2025 due to CPEC*[3] *projects is considered in this scenario* (NTDC, 2018).

(v) *The ARE policy of 2019s target of 20% RE generation in the supply mix by 2025 is also accounted for in the model.*

(vi) *The GDP of Pakistan is taken as an external variable and the World Bank database is used.*

---

[1] T & D: Transmission and Distribution.

[2] NTDC: National Transmission and Distribution Company.

[3] CPEC: China Pakistan Economic Corridor.

Table 7.1 Power plants data

Power plant technology	Average life (years)	Generation (GWh)	Capacity (MW)[a]	Capacity factor (%)[a]	Utilization factor (%)[a]	CO_2 emissions (g CO_2/kWh)
Hydro	80	28,707	6480	98	46	12.8[a]
Oil[b]	30	30,818	8027	63	58	1502.86[b]
Coal[c]	30	136	50	80	23	790[c]
Gas[d]	30	32,923	40,970	85	60	531[d]
Nuclear	40	3077	462	79	85	12.51
Wind	20	0	0	33	22	11.38
Solar	25	0	0	35	29	11.38

Source Qudrat-Ullah (2022)
[a]Capacity factor and utilization factor are average values based on the data of all the plants in operation in Pakistan (for wind and solar the source is (Nicholas & Buckley, 2018); the majority of hydro generation is reservoir-based; [b]majority of oil is furnace oil (imported) and large transportation distances are the factor in the higher emissions; [c]majority of coal plants use locally available coal (transportation need is little); [d]majority of gas-based plants are single-cycle plants and use natural gas. Currently, nuclear waste is kept in pools at the site of the power plants in Pakistan, a relatively cheaper and low CO_2 emitting operation (WNA, 2020)

Table 7.2 Cost of generation technologies

Power plant technology	Capital cost[a] ($/kW)	Fixed O&M cost ($/kW-year)	Variable O&M cost ($/MWh)	Key parametric values and assumptions
Hydro[b]	680	19.8	0.0	• RE and nuclear technologies' data is from IRENA • The Discount factor is 10% for the the2008–2035 period • Data on coal, oil, and gas power plants are based on IEA,[4] PEY19,[5] and the least values are used to reflect domestic distances (e.g., labor cost, transportation is cheaper)
Coal	1072	40.41	2.43[f]	
Gas[c]	536	54.4	14.04	
Oil	710	6.97	4.48	
Nuclear	2805	30	0.51	
Solar	618	39.56	0.0	
Wind[d]	1039	26.22	0.0	

Source Qudrat-Ullah (2022)
[a]These are levalised costs (Timilsina, 2020); [b]Reservoir-based hydropower data is used; [c]Majority of gas-based plants in Pakistan are single cycle; [d]Pakistan have predominantly onshore wind power plants; US$ are converted to Pakistani Rupees using an annual exchange rate as is reported by State Bank of Pakistan and vice versa (e.g., US$ = 59.86 in the year 2005–2006); [f]Most of the coal used is local and costs less due to shorter distances and cheap labor (Fatima & Nasim, 2019; Jahangir et al., 2021).

(vii) *Cost data for generation technologies are based on levalized costs.*

[4] IEA: International Energy Agency.
[5] PEY19: Pakistan Energy Yearbook 2019.

(viii) *As per NTDC's projections, a carbon tax of US$ 31/tCO_2 is considered effective from July 1, 2025* (NTDC, 2018).

(ix) *The annual exchange rate as published by the State Bank of Pakistan is used to cover US$ into Pakistan Rupees and (vice versa) and US$ is indexed to the base year.*

7.2.2 Indigenous-Resource-Intensive Low-Carbon (ILC) Scenario

In contrast to BAU, the reference scenario, which is based on Pakistan's current conditions where the energy policies continue to provide fiscal incentives for electricity generation without any preference for any type of technology (with IPPs are dominantly investing in pro-oil and pro-coal generation), the ILC scenario emphasizes the use of available renewable and cleaner sources of energy. Specifically, it focuses on:

(i) the exploitation of indigenously rich hydro resources,
(ii) abundantly available other renewables (i.e., solar and wind), and
(iii) low-carbon producing nuclear generation.

It should be noted that Pakistan is rich not only in uranium deposits but also in having the indigenous capability of nuclear fuel fabrication. It has achieved self-sufficiency in the nuclear fuel cycle, i.e., mining of uranium, making of yellow-cake (U3O8), and fabrication of ready-to-use fuel since 1979 (WNA, 2014). In this scenario, thermal power is bearing additional costs due to the "environmental emissions tax". An emission tax rate of $31/ton of CO_2, found through a detailed sensitivity analysis, is applied in the BAU scenario. Such a tax is considered a step toward a much-needed low-carbon economy (Mahmood et al., 2020).

7.2.3 Nuclear Power Dominant (NPD) Scenario

In this progressive, intensive production of clean energy scenario, we consider 30% nuclear power by 2030 in the electricity supply mix of Pakistan. This nuclear power dominant scenario is similar to the ILC scenario in the utilization of indigenous hydro and renewable energy resources, emphasizing the progressive development of nuclear power in Pakistan. Consideration of 30% nuclear power by 2030 is based on the ambitious expansion plan of the Government of Pakistan. Ina recent empirical study, Mahmood et al. (2020) found that nuclear power, with improved safety measures, is the way forward for Pakistan to control power-related environmental emissions. We expect the implementation of such a scenario will help the policymakers in Pakistan to realize the goal of having a low-carbon economy and met the targets of CO_2 reductions.

7.3 Policy Assessment: Evaluation of BAU Scenario

The BAU scenario represents the conditions and incentives of the current Energy Policy of Pakistan. Utilizing our validated revised MDESRAP, here we will present a systematic and multidimensional assessment: to what extent this policy provides affordable, reliable, and cleaner production of electricity. We will specifically assess if the current energy policy of Pakistan will achieve the target reduction of electricity-related CO_2 emissions.

7.3.1 Dynamics of Electricity Supply Mix

Based on reference data of Pakistan's case, Energy Policy incentives, and INDC's mechanisms, Fig. 7.2 displays the MDESRAP's outcome-based total electricity generation in Pakistan for the period of study from 2008 to 2035. The share of technology in the investments is dependent on the relative cost of the technology and the effect of the NPV-of-investment-premium on investments. Consequently, the technology with relatively lower costs and a higher (reward) premium is substituted. However, the substitution process is delayed because it takes time to refurbish the old capacity. Thus, the relatively more acquisition of the production capital in the last decade of the time horizon causes an increase, relative to the pre-policy case, in the electricity supply. Post-2020, there appears at least four major developments:

(i) the dominant role of IPPs in thermal power generation,
(ii) the dominant presence of CPEC-based coal-fired power plants,
(iii) consistent with Pakistan's INDC-based plans, hydropower is steadily increasing its share in the electricity supply mix, and
(iv) the share of low-carbon technologies (i.e., nuclear, solar, and wind) is increasing at a relatively slower pace.

We can observe in Fig. 7.2, during the post-2020 era, a major shift in electricity generation toward CPEC-based coal-fired power plants. On the other hand, the clean sources of electricity production including hydro, nuclear, solar, and wind are increasing their share in the electricity mix at a much slower rate. Only a strong focus on cleaner production and a quick re-adjustment in the Energy Policy of Pakistan, for example through the implementation of the planned CO_2 tax regime, can support relatively faster penetration of clean energy-producing technologies into the electricity supply mix of Pakistan.

In the case of hydro-based generation, where Pakistan has rich domestic resources, there is a rewarding premium[6] as shown in Table 7.2 that favors the choice of cleaner-generation technology. However, the unit cost seems to surmount the effect of this premium. Figure 7.2 portrays the switch from hydropower plants as a result of

[6] These reward/penalty premiums are based on our MDESRAP-based analysis with a carbon-tax regime in place.

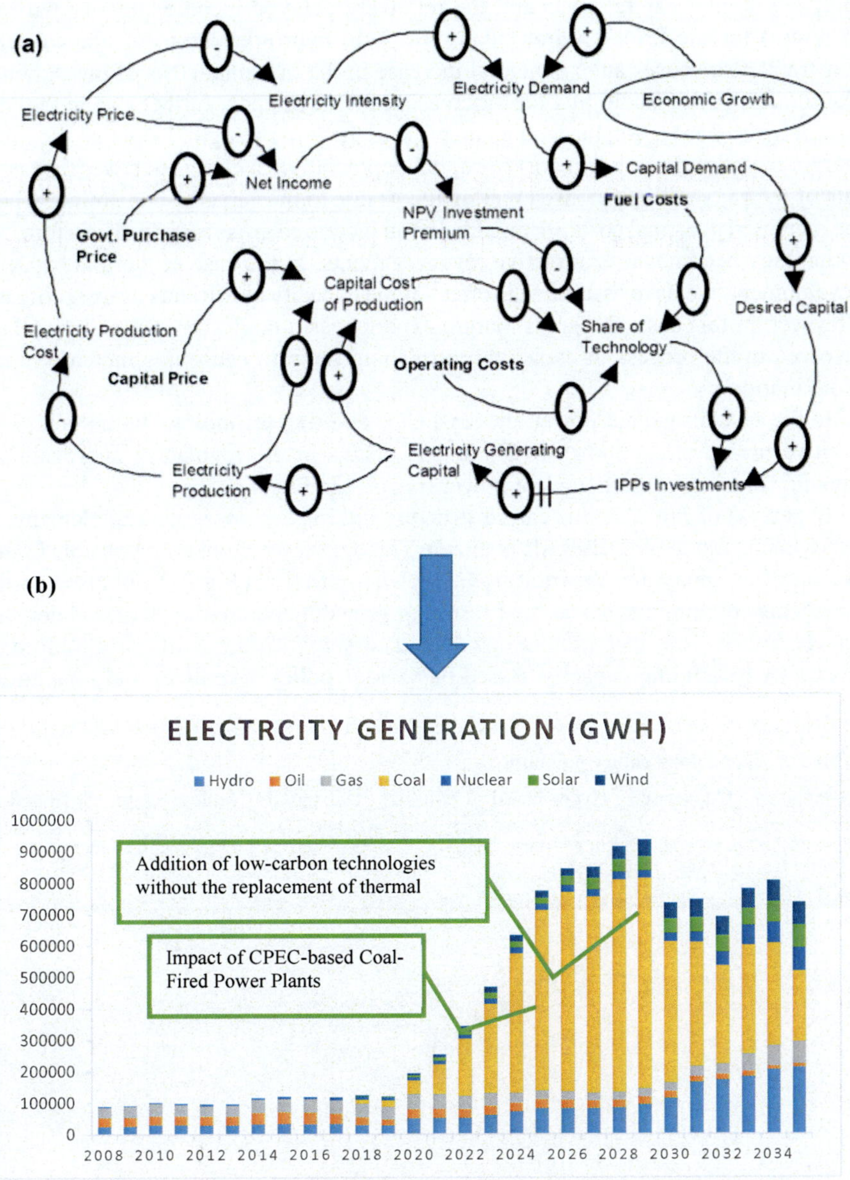

Fig. 7.2 Dynamics of electricity generation in Pakistan for the period 2008–2035

relatively higher capital costs and the relatively delayed income stream (real rate of return) for the IPPs (Qudrat-Ullah, 2015). In hydro-development, the capacity constraint also causes an exponential increase in the marginal costs of investments. Consequently, this additional cost-increasing factor brings a further rise in the unit cost of hydro generation through an increase in the gross margin. Note that the gross margin includes the per-unit cost of capital depreciation, a charge to cover the target rate of return, and the increased marginal cost due to the capacity constraints. Thus, the increased marginal costs are reflected as an increased gross margin. Therefore, the technology becomes less attractive for investments. In the case of thermal capacity development, we have assumed the effect of the capacity constraints as insignificant. However, in this case, the gross margin experiences a modest increase. The delays involved in the process of capacity refurbishing seem to cause this increase in the gross margin.

In the case of nuclear power, the least CO_2 emitting technology with the highest positive premium, as shown in Table 7.3, delays in the regulatory approvals are causing slower growth in nuclear power (please see Fig. 7.3).

In part (a) of Fig. 7.2, this causal structure linking various structural elements of our dynamic model, MDESRAP, is shown. This causal structure is responsible for the outcome (i.e., generated electricity) as is shown in part (b) of Fig. 7.2. For instance, the increasing economic activities and resulting growth generate the increased demand for electricity. This increased demand sends a signal to the market for additional electricity-generating capacity. Based on various policy incentives and regulations

Table 7.3 Technology choice premium

Technology	Coal-based	Hydro-based	Oil-based	Gas-based	Nuclear-based	Solar and wind-based
Premium ($/MWh):	171	−45	98	−26	−56	−39

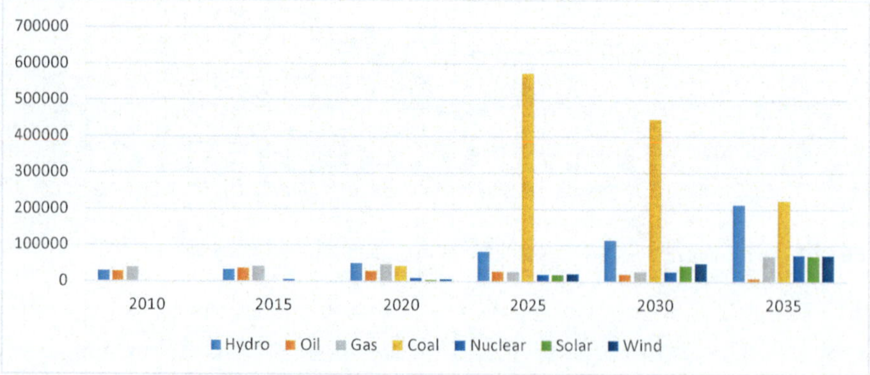

Fig. 7.3 Dynamics of electricity supply mix

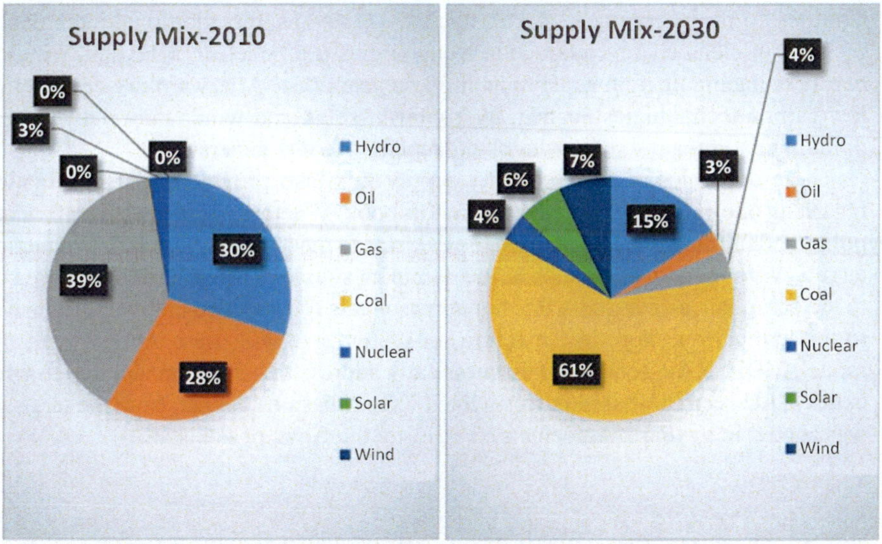

Fig. 7.4 Electricity supply mix of Pakistan for 2010 and 2035

(e.g., Govt.'s power purchase agreements, capacity payments, tax incentives, etc.,), the IPPs' investments bring in new capital (i.e., electricity-generating power plants), and thereby electricity is generated.

In addition, we can see the economic pressure behind the so-called "dash for gas" and the fast income generator "CPEC-based coal-fired power plants" which characterized the capacity-building intentions[7] of most of IPPs during the early years of policy promulgation.

Overall, when we look at Fig. 7.4, which provides a percentage-wise composition of power generation technologies in the year 2010[8] versus the year 2030, there is a clear dominance of thermal power plants in the electricity supply mix of Pakistan. While in the year 2010 both oil-fired and gas-fired power plants had the major share in the supply mix. However, in the year 2030 period, most of the electricity in Pakistan was generated by coal-fired power plants. The dominance of coal-fired power plants is consistent with the government's ambitious plan to utilize the abundantly available local resource (e.g., Thar coal deposits) together with the commissioning of several CPEC-based coal-fired power plants (Qudrat-Ullah, 2022). The coal-based generation remains the dominant source of electricity supply from late 2021 to the end of the year 2030, as can be seen in Figs. 7.2 and 7.3. Only after 2030, the year of compliance for the reduction of CO_2 emissions set by the Paris Agreement (PA), effective replacement of thermal generation by low-carbon technologies occurs.

---

[7] By the end of 2006, the government of Pakistan had received investment offers from IPPs of about 4 GW capacity of which 2.5 GW was for gas-fired power plants, 1.2 GW for oil-fired power plants, and the rest (0.3 GW) for the coal-fired power plants.

[8] For a comparable view of supply mix, we have used the year 2010 (instead of our base year, 2008).

Contrary to the government of Pakistan's expectations, the additional incremental increase in clean energy production technologies (i.e., nuclear, solar, and wind) is hardly replacing thermal generation until the year 2031. Afterward, we can see that low-carbon technologies nuclear, hydropower, solar, and wind, and relatively less polluting natural gas-fired power plants replace thermal generation.

These dynamics of the electricity supply mix also provide initial information regarding one of the key objectives set in this book: Whether or not the current energy policy and initiatives in Pakistan enable Pakistan to meet the PA-based CO_2 reduction targets. With thermal generation as the dominant source of electricity supply even up to the compliance year-2030, it is not unreasonable to expect an unfavorable answer to this critical question. In the following section, we will use hard data (i.e., our model, MDESRAP-generated) to empirically address this question and find to what extent BAU scenario-based power-related CO_2 emissions deviate from the target as was agreed in the Paris Agreement in 2015 by the Govt. of Pakistan.

7.3.2 Does the Current Policy of Pakistan Meet or Miss the PA Targets of CO_2?

As we saw in the previous section about the dynamics of the electricity supply mix of Pakistan, thermal generation—predominantly with CPEC-based coal-fired power plants—is still the dominant source for the period from the year 2010 to the year 2035. Table 7.4 list the CO_2 emissions quotients associated with various technologies. Although nuclear and renewable technologies do not emit CO_2, these accounted CO_2 emissions are based on a life-cycle approach assessment (e.g., CO_2 emission during, mining, transportation, and fabrication of uranium and the transportation and installation of machinery for renewable technologies. These CO_2 emissions quotients are in the calculation of CO_2 emissions by each of the electricity-generation technologies in all the scenarios reported in this book. In Table 7.5, we present the summary of MDESRAP-generated power-related CO_2 emissions in Pakistan and the deviation from the PA target (i.e., 64.3 $MtCO_2$ emissions to be met by the year 2030).

Contrary to the expectation of Govt. of Pakistan, their voluntarily made commitment to the reduction of CO_2 emissions is not going to be achieved by the current Energy Policy of Pakistan. On the other hand, the current energy policies and mechanisms (i.e. BAU and CPEC-based projects), appear to deviate from the power-related CO_2 emissions by the addition of 347.3 $MtCO_2$ to the target of 64.3 $MtCO_2$ in the year 2030. Consistent with the results of earlier studies (e.g., Qudrat-Ullah, 2022), staying the course is hardly a viable option for Pakistan. The situation of CO_2 emissions in the year 2035 is even the worst and most bleak primarily due to the dominance of CPEC-based coal-fired power plants. Therefore, with a relatively shorter time window left before the compliance year 2030, Pakistan has to make some critical energy policy re-adjustments decisions to bring additional clean power-producing technologies to the electricity supply mix of Pakistan.

Table 7.4 CO_2 emissions quotients for various technologies

Technology:	Coal-based	Hydro-based	Oil-based	Gas-based	Nuclear-based	Solar-based	Wind-based
CO_2 Emissions (g/kWh):	986	249.4	670.6	249.4	25	372	65.5

Source EPA (2021)

Table 7.5 Summary of CO_2 emissions contribution by the generation technologies

Year	Power-related emissions ($MtCO_2$)	Contributing technologies		Target deviations ($MtCO_2$)
		Thermal (i.e., oil, coal, and gas) (%)	Low carbon (i.e., hydro, nuclear, solar, and wind) (%)	
2010	73.3	99.3	0.7	–
2015	89.1	99.2	0.8	–
2020	112.8	99.4	0.6	–
2025	517.5	99.5	0.5	–
2030	**411.7**	99.3	0.7	**+347.3**
2035	2095.4	99.4	0.6	–

Table 7.6 Comparison of electricity-related CO_2 emissions in Pakistan

Scenario	Cumulative CO_2	Escaped CO_2
Pre-policy (RE2006) scenario ($MtCO_2$[9])	277.3	1321.5
Post-policy-BAU scenario ($MtCO_2$)	411.7	1623.4
% change	+49	+23

It should also be noted that the reference scenario, BAU, represents a scenario of unconstrained CO_2 emissions because Pakistan's Energy Policy does not include any environmental constraints or limits. The cumulative CO_2 emissions reflect a mix of coal, oil, gas, hydro, nuclear, solar, and wind power production capacities. Despite the introduction of the RE-2006 policy incentives, the increasing trend in CO_2 emissions is further reinforced. This corresponds to the intention of IPPs to preferably invest in thermal generation, especially in coal-fired power plants (Qudrat-Ullah, 2015). Under the influence of policy incentives, the cumulative CO_2 emissions (inside the national boundaries of the country) attain a level that is 48% above the pre-policy case (please see Table 7.6). Similarly, there is a substantial increase (of 23%) in CO_2 (outside the national boundary of the country). Thus, the overall impact of the Energy Policy incentives on the environment is not friendly.

7.3.3 A Look at the Power-Related CO_2 Emissions in the PIS Region

For a comparative purpose, the projected power-related CO_2 emissions of the PIS region countries are listed in Table 7.7. Although each country is diverse in urbaniza-

[9] $MtCO_2$: Million tons of Carbon Dioxide.

Table 7.7 Projected power-related CO_2 emission ($MtCO_2$) in the PIS region

PIS region country	Annual rate (%)	2030 (Projected)	Source
Pakistan	11.3	1321.5	This study
India	11.7	1623.4	IEA (2021)
KSA[10]	11	670.8	Rahman et al. (2021)

tion, population, and economic activities, the annual rate of emission CO_2 emissions from existing and new power infrastructure are about the same, i.e., close to 11%. Both Saudi Arabia and India have stated national policies to implement CO reduction measures to achieve the PA-based targets by 2030 (CAT, 2021; IEA, 2021). Given the slower implementation (e.g., in March 2021), Saudi Arabia announced the Saudi Green Initiative which includes the target to increase renewables in the primary energy mix to 50% by 2030, from less than 0.1% in 2020 is progressing very slowly (Rahman et al., 2021). The need for a systematic system-wide assessment of electricity-related CO_2 emissions becomes even more pronounced. As the focus of this book is Pakistan's case, in the following sections we will explore the potential pathways for Pakistan to meet its PA-based targets and move towards achieving a low-carbon economy. We will present the results of the evaluation of the two clean-energy-focused scenarios, i.e., ILC and NPD scenarios.

7.3.4 The Dynamics of Electricity Prices in the BAU Scenario

As planned and unplanned blackouts and outages of electricity in Pakistan are an ongoing drag on the well-being of people and sustained development of economic activities, the availability of electricity, therefore, is of primal concern for policymakers in Pakistan. However, the availability of electricity alone is not enough to lessons people's sufferings and energize economic and entrepreneurial activities in the country, the price of electricity, the affordability aspect of electricity have to be ensured.

As previously discussed in the case of electricity supply mix and CO_2 emissions, continuing with the existing energy policies and mechanisms can severely affect the consumers of electricity on the affordability front—as well. Figure 7.5 shows the dynamics of electricity prices for the duration of this study. In the post-2019 era, it appears that two important developments regarding electricity prices in Pakistan are at play: (i) the use of local coal (e.g., Thar coal), and (ii) the development of imported-fuel-based coal-fired power plants under the CPEC-based projects (Qudrat-Ullah, 2022). Relatively lower prices up to the year 2025 can be ascribed to the use of the locally available and cheaper supply of coal. However, after the year 2024, this

[10] KSA: Kingdom of Saudi Arabia.

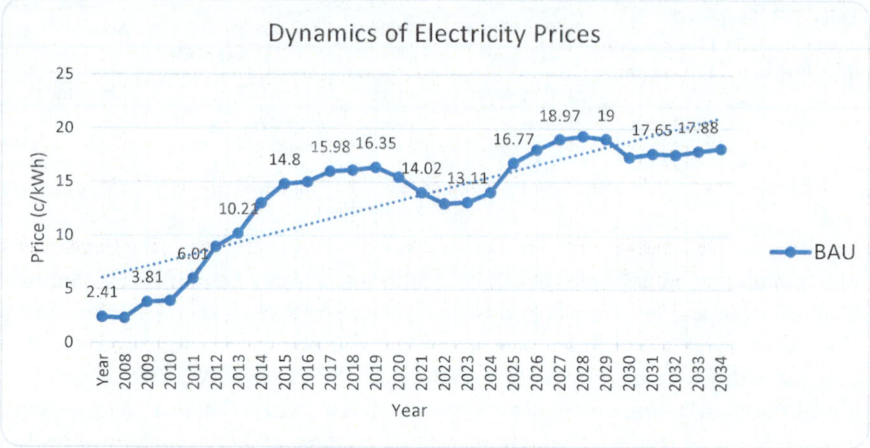

Fig. 7.5 Dynamics of electricity supply for the period from 2010 to 2035

effect of cheaper coal appears to be offset by the use of imported coal for CPEC-based coal-powered plants. Only after the year 2029, we can see the relative decline in the electricity prices due to the relatively increased share of hydro, nuclear, solar, and wind up to the end of the year, 2035. Overall, this trend shows that consumers will be paying higher electricity for the near future until the current energy policy of Pakistan is not adjusted toward the use of low-carbon and domestic resource-based technologies.

There is another aspect of the rising electricity prices in Pakistan. For oil-fired power plants especially from 2008 to 2025, the fuel requirement of oil-based generation (the other IPPs preferred technology for investment) is met through imports. Being the less capital-intensive technology and an earlier income stream generation potential drive the increased IPPs investment in the oil-based power plants. Consequently, more oil is imported. This scenario explains how the substantial amount of hard-earned foreign exchange of the economy of Pakistan is being spent to keep the oil-based power plants running. In the post-2020 era, instead of replacing oil-fired power plants with equally burdensome (e.g., CPEC-based power plants are run on imported coal) and highly pollutant coal-fired power plants, efforts should be channeled toward cleaner electricity-generating technologies—Without such re-adjustments, it appears unlikely to achieve a low-carbon economy in Pakistan.

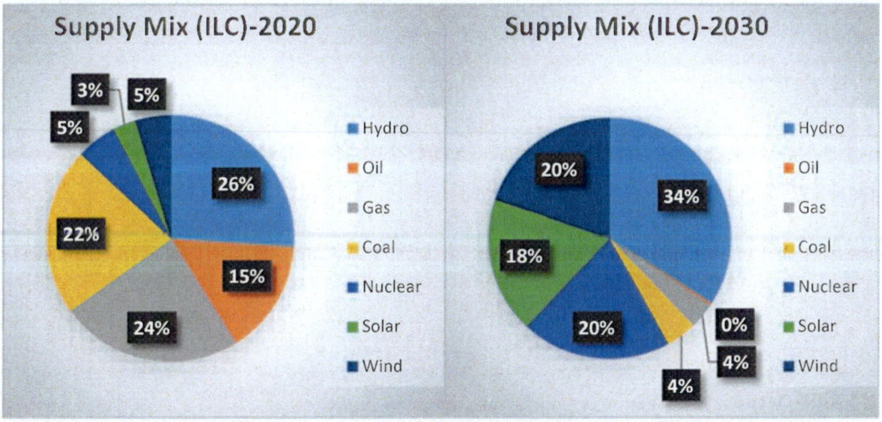

Fig. 7.6 Dynamics of electricity supply mix-percentagewise

7.4 Dynamics of Indigenous-Resource-Intensive Low-Carbon (ILC) Scenario

After achieving the first key objective of this book: To assess the current Energy Policy of Pakistan to see if the PA-based CO_2 emissions reduction targets are on point or not. In the previous section (i.e., 7.2), it was shown that staying the course is hardly an option for the energy policymakers in Pakistan. Now, to meet the other objective: What is the best electricity supply mix for Pakistan to achieve affordable and cleaner production of electricity? Here, we evaluate the first option: the ILC scenario. The exploitation and use of local resources for the cleaner production of electricity is the major feature of this scenario. We fixed the total supply of electricity equal to that in the BAU scenario and ran various MDESRAP-based simulations to find the appropriate supply mix.

7.4.1 Dynamics of Electricity Supply Mix and CO_2 Emissions in the ILC Scenario

Figures 7.6 and 7.7 display the dynamics of the electricity supply mix under ILC. The same amount of electricity is generated but with a different supply mix. In this clean-energy-focused scenario, in the year 2030, a PA-based targets compliance year, a substantial amount of electricity (i.e., 92%) is produced by clean-energy technologies. The remaining 8% of electricity is produced by coal-fired[11] (4%) and gas-fired (4%) power plants.

[11] In this scenario, it is assumed that the CPEC-based coal-fired power plants will implement clean-coal technologies that will result in relatively lower CO_2 emissions.

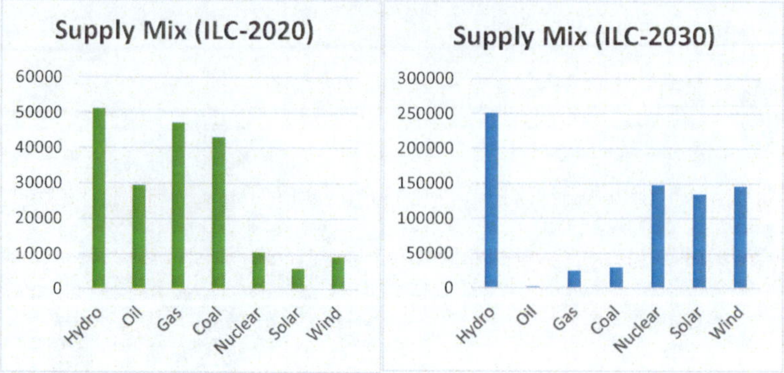

Fig. 7.7 Dynamics of electricity supply mix-quantity wise

In this alternate scenario, ILC, the hydropower, with which Pakistan is endowed with rich domestic sources for large dams (e.g., Dasu dam), has the major (i.e., 34%) share in the electricity supply mix of Pakistan. This development is in sharp contrast to the case of our reference scenario, BAU, where coal-fired power plants had the largest share in the electricity supply mix of Pakistan, especially for the period from the year 2020 to the year 2030. Solar, wind, and nuclear, all three clean energy-generation technologies gain a substantial share in the electricity supply mix (i.e., solar power = 18%, nuclear power = 20%, wind power = 20%) during the period from the year 2020 to the year 2030. It is very encouraging to see the increasing role of clean energy-generating technologies based on the local availability of the resources and capability (e.g., to manufacture the fuel for nuclear power plants from the locally available uranium deposits (e.g., in D.G. Khan mines)).

7.4.2 Does the ILC Scenario Meet or Miss the PA Targets of CO_2?

A closer look at Figs. 7.6 and 7.7, can visually show that the ILC scenario meets the PA-based CO_2 reduction targets of Pakistan. In Table 7.8, we present a comparative view of electricity-related CO_2 emissions in both the BAU and ILC scenarios

Table 7.8 Electricity-related CO_2 emissions contribution by BAU and ILC scenarios

Year	Power-related emissions ($MtCO_2$)				Target[a] deviations ($MtCO_2$)
Scenario	2020	2025	**2030**[a]	2035	
BAU	112.8	517.5	**409.9**	2095.4	**+347.3**
ILC	113.3	367.6	**50.4**	35.7	**−13.9**

[a]Target = 64.3 $MtCO_2$ by 2030

where it is quantitatively shown, as produced by MDESRAP-based analysis, that the ILC convincingly achieves the target of CO_2 reduction by a margin of 13.9 $MtCO_2$. The primary reason behind these lower CO_2 emissions is the desired substitution of cleaner energy-generating technologies in place of the dominant thermal generation of electricity in Pakistan, especially during the post-2023 period. In particular, the share of nuclear power increased from 5% in 2020 to about 20% in 2030. Likewise, the share of solar power and wind power witnessed an increase of about 20 and 33%, respectively. For this substantial progress in the capacity of cleaner electricity-generating technologies in the electricity supply mix of Pakistan, the ongoing collaboration with China in nuclear power and falling prices of solar PV and wind technologies in the global market appears to be the major support factors (Nicholas & Buckley, 2018; Timilsina, 2020). Therefore, it is not unreasonable to assert that this alternate scenario, ILC, not only helps Pakistan to meet its PA-based CO_2 reduction commitments but also puts it on track to achieving a low-carbon economy.

Overall, it should be noted the increasing share of clean energy-generating technologies in the ILC scenario is in line with the Government of Pakistan's various stated policy incentives and initiatives. However, to realize these outcomes there is a need to reinforce the efforts and embrace some adjustments in the policy incentives to spur the implantation of clean energy-generating technologies and projects in the country (Mengal et al., 2019; NTDC, 2018). Successful implementation of ILC like scenarios is one of the viable ways to achieve the desired reduction in electricity-related CO_2 emissions and move towards a low-carbon economic regime.

7.4.3 Dynamics of Electricity Prices in the ILC Scenario

The price of electricity directly relates to the affordability aspect of electricity use by the consumers. Figure 7.8 shows the evolution of the electricity prices in the ILC

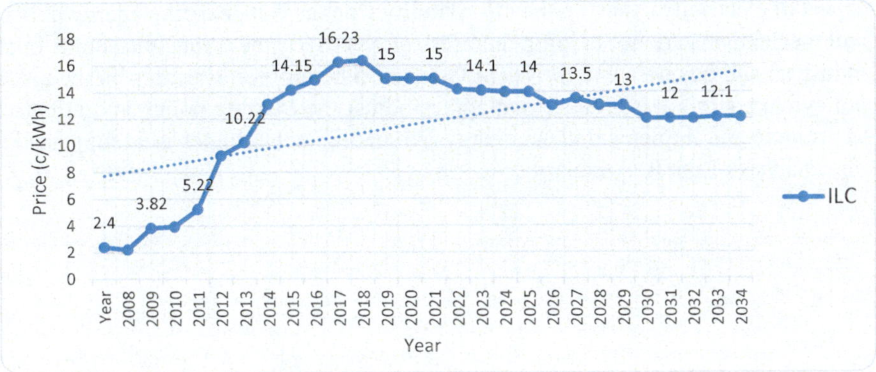

Fig. 7.8 Dynamics of electricity prices in the ILC scenario

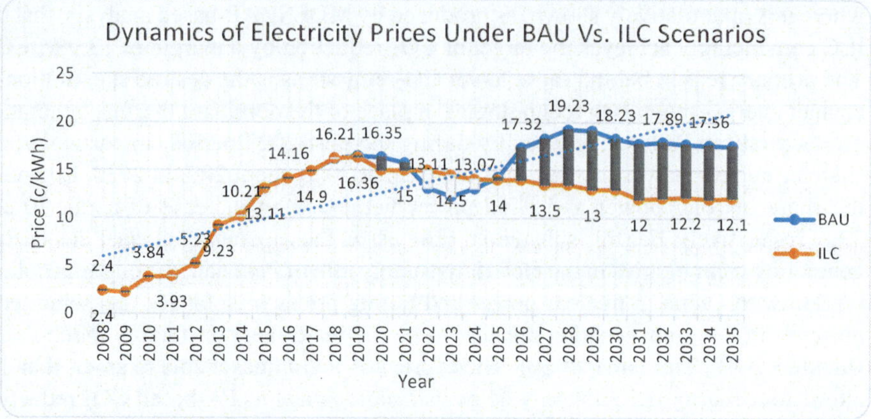

Fig. 7.9 Comparative view of electricity prices in both scenarios

scenario. To see how the ILC fares on the affordability front compared with the BAU scenario. Figure 7.9 depicts the dynamics of the electricity prices under bot both scenarios.

Regarding the affordability aspect of electricity supply in Fig. 7.9, we can see the relatively lower electricity prices during the period from 2021 to the beginning of the year 2025 under the BAU scenario. However, afterward, as compared with the BAU scenario, there are substantially lower prices in the ILC scenario owing to a favorable premium for cleaner production (i.e., when a carbon tax of US$31 is implemented, it penalizes the thermal generation and rewards the cleaner generation). Due to the relatively faster (as compared with the case of the BAU scenario) replacement of oil and coal-based power capacity by low carbon technologies that have the least operating cost, the electricity prices continue to fall up to the end of the study period, 2035. On average, for the period from the year 2020 to the year 2035, consumers of electricity will save 20% annually (16.65 versus 13.38), a significant factor in the affordability aspect of electricity. Also, on the affordability front as well, existing energy policies and mechanisms are not helping the users of electricity, be it the households or the industrial and business enterprises in Pakistan. Therefore, it warrants on the part of policymakers in Pakistan to revisit and re-adjust the existing policy and initiatives to promote and implement clean energy-generating technologies as is suggested by this analysis of the ILC scenario.

7.5 Dynamics of Nuclear Power Dominant (NPD) Scenario

Regarding the positive effects of nuclear power, we have seen in the previous section the evaluation of our ILC scenario that with a 20% share of nuclear power in the electricity supply mix of Pakistan not only Pakistan can meet its PA-based CO_2 reduction commitments but also can begin the journey towards a low-carbon economy. Building on this empirical finding and keeping an eye on the major objective of this book: What is the best electricity supply mix for Pakistan to achieve affordable and cleaner production of electricity? Here, we evaluate our second option: the NPD scenario. In addition to the major feature of the ILC scenario: the exploitation and use of local resources for the cleaner production of electricity, we are adding a specific constraint to our dynamic model MDESRAP—"ensure at least 30% of nuclear power in the best electricity supply mix for Pakistan." As was the case in the ILC scenario, we fixed the total supply of electricity equal to that in the BAU scenario and ran various MDESRAP-based simulations to find the appropriate electricity supply mix for the electricity sector of Pakistan.

7.5.1 Dynamics of Electricity Supply Mix and CO_2 Emissions in the NPD Scenario

The electricity supply mix of Pakistan under the NPD scenario is shown in Figs. 7.10 and 7.11. The same amount of electricity is generated but with a different supply, mix where the dominant share is captured by nuclear power, as was expected. In this renewable-energy and nuclear power-focused scenario, in the year 2030, a PA-based targets compliance year, a substantial amount of electricity (i.e., 94%) is produced

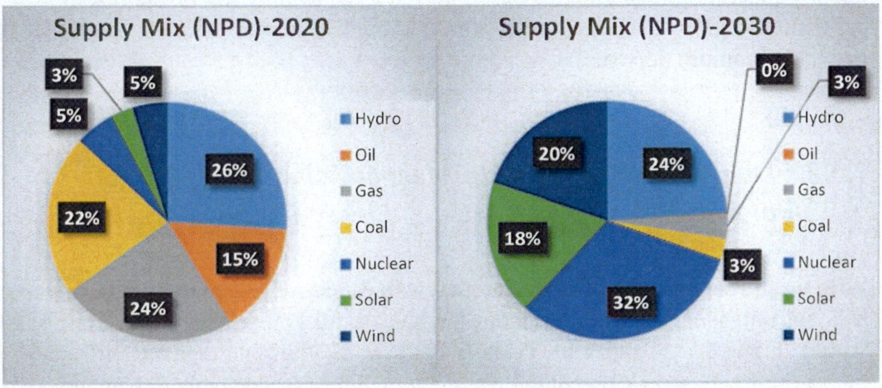

Fig. 7.10 Dynamics of electricity supply mix-percentagewise in NPD scenario

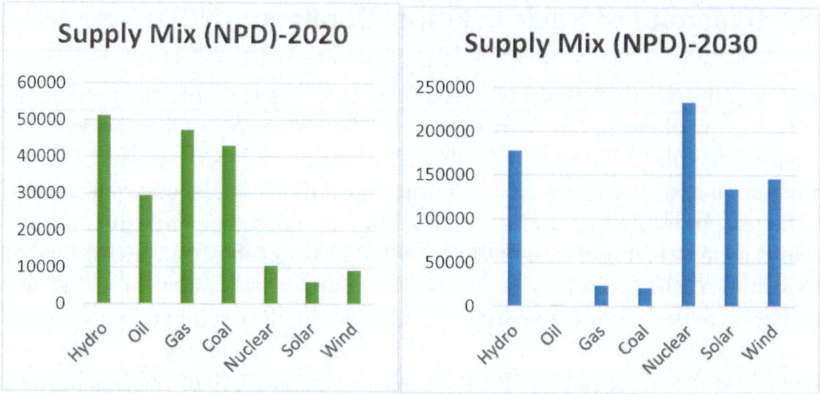

Fig. 7.11 Dynamics of electricity supply mix-quantity wise in NPD scenario

by clean-energy technologies. The remaining 6% of electricity is produced by coal-fired[12] (3%) and gas-fired (3%) power plants.

In this 2nd alternate scenario, NPD, the nuclear power, with which Pakistan is endowed with abundant uranium deposits and self-sufficient capability for nuclear fuel fabrication and operations of nuclear power plants (Qudrat-Ullah, 2022), has the major (i.e., 32%) share in the electricity supply mix of Pakistan. This development of dominance of clean electricity is in sharp contrast to the case of our reference scenario, BAU, where coal-fired power plants had the largest share in the electricity supply mix of Pakistan, specifically for the period from the year 2020 to the year 2030. Solar, wind, and hydro, all three clean and renewable energy-generation technologies, also gain a substantial share in the electricity supply mix (i.e., hydro = 24%, wind power = 20%, and solar power = 18%) during the period from the year 2020 to the year 2030. It is very encouraging to see the increasing role of clean and renewable energy-generating technologies based on the local availability of the resources and capability (e.g., to manufacture the fuel for nuclear power plants from the locally available uranium deposits).

7.5.2 *Does the NPD Scenario Meet or Miss the PA Targets of CO_2?*

With the dominant share of clean and renewable electricity-generating technologies, it is not unreasonable to think about the successful compliance of CO_2 emissions targets. Figures 7.10 and 7.11, visually confirm the thinking that the NPD scenario meets the PA-based CO_2 reduction targets of Pakistan. In Table 7.9, we present a

[12] In this scenario, it is assumed that the CPEC-based coal-fired power plants will implement clean-coal technologies that will result in relatively lower CO_2 emissions.

Table 7.9 Electricity-related CO_2 emissions contribution by BAU and NPD scenarios

Year	Power-related emissions ($MtCO_2$)				Target[a] deviations ($MtCO_2$)
Scenario	2020	2025	**2030**[a]	2035	
BAU	112.8	517.5	**409.9**	2095.4	**+347.3**
NPD	112.9	365.2	**43.3**	17.3	**−21**

[a]Target = 64.3 $MtCO_2$ by 2030

comparative view of electricity-related CO_2 emissions in both the BAU and NPD scenarios where it is quantitatively shown, as produced by MDESRAP-based analysis, that the NPD convincingly achieves the target of CO_2 reduction by a margin of 21 $MtCO_2$. The primary reason behind these lower CO_2 emissions is the desired substitution of cleaner and renewable energy-generating technologies in place of the dominant thermal generation of electricity in Pakistan, especially during the post-2023 period. Both gas and coal-based electricity generation capacity is on a diminishing track under the NPD scenario. Therefore, it is not unreasonable to assert that this alternate scenario, NPD, not only helps Pakistan to meet its PA-based CO_2 reduction commitments but also utilizes the local resources and capabilities to put it on the track to achieving a low-carbon economy.

Overall, it should be noted that the dominant share of clean energy-generating technology (i.e., nuclear power) in the NPD scenario is in line with the Government of Pakistan's ambitious nuclear-power expansion plan (NTDC, 2018). However, to realize these outcomes of the NPD scenario, there is a need to reinforce the efforts and embrace some adjustments in the policy incentives to spur the implementation of clean and renewable energy-generating technologies in the country (Mengal et al., 2019; NTDC, 2018). Successful implementation of NPD-like scenarios is one of the viable ways for Pakistan to achieve the target reduction in electricity-related CO_2 emissions and a low-carbon economy smoothly.

7.5.3 Dynamics of Electricity Prices in the NPD Scenario

The price of electricity directly relates to the affordability aspect of electricity use by the consumers. Figure 7.12 shows the evolution of the electricity prices in the NPD scenario. To see how the INPD fares on the affordability front compared with the BAU scenario, Fig. 7.13 depicts the dynamics of the electricity prices under both scenarios.

Compared with the BAU scenario where the electricity prices for relatively lower during the period from 2021 to the beginning of the year 2026, the NPD scenario shows higher prices. These higher prices reflect the capital-intensive nature of investments in nuclear power technology. However, when newly added nuclear power plants become operational, there are substantially lower prices in the NPD scenario. The planned implementation of a carbon tax that is imposed on CO_2 emitting generation technologies appears to make nuclear power price competitive with thermal

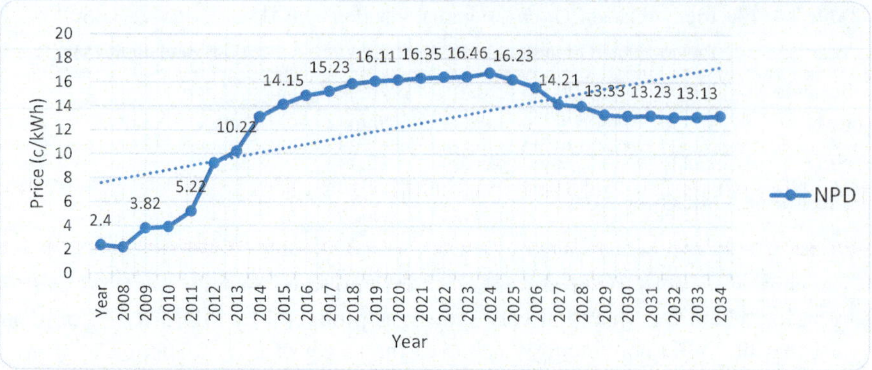

Fig. 7.12 Dynamics of electricity prices in the NPD scenario

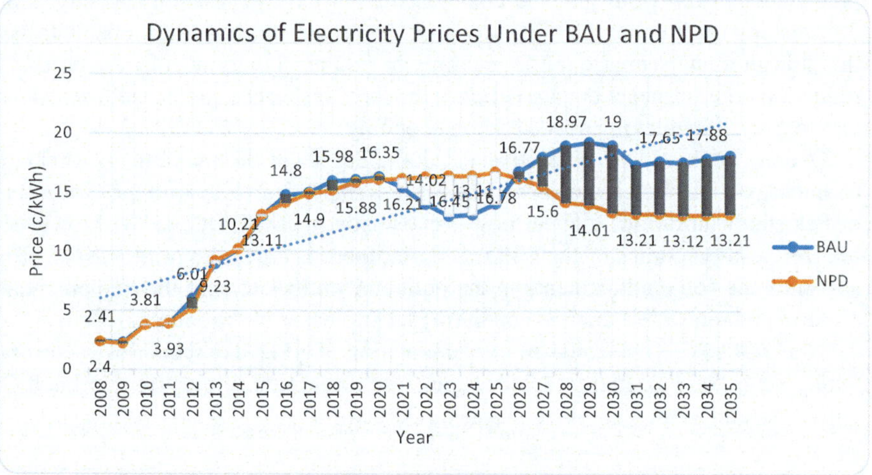

Fig. 7.13 Comparative view of electricity prices in BAU and NPD scenarios

generation-the dominant electricity producer in the BAU scenario. As the more operational capacity of nuclear power is added, especially from the year 2026 to the year 2033 when thermal generation is effectively replaced by cleaner and renewable electricity generation technologies, the price of electricity in the NPD scenario becomes even cheaper. These relatively lower prices of electricity in the NPD scenario continue to fall up to the end of the study period, 2035.

On average, for the period from the year 2020 to the year 2035, consumers of electricity will save 11% annually (16.6 versus 14.8), a significant factor in the savings for the majority of the low-income strata of the population of Pakistan, the affordability aspect of electricity. With a dominant share of clean generation (i.e., with a 32% share of nuclear power) in the electricity supply mix of Pakistan, these competitive prices in the NPD scenario warrant the part of policymakers in Pakistan to revisit and re-adjust the policy investments to promote and implement more clean energy-generating nuclear power plants in the country.

7.6 In Search of the Best Policy Scenario for the Reduction of CO_2

Designing policy incentives and initiatives for achieving the objective of supplying reliable, affordable, and cleaner electricity is a complex and difficult task at best. However, with the help of our dynamic model, MDESRAP, we have assessed the existing policy scenario, BAU, and have evaluated two alternate scenarios, the ILC and the NPD scenarios. In the post-2022 era, the availability of electricity supply in Pakistan is expected to exceed its annual demand (e.g., due to new CPEC-based coal-fired power plants). However, at the outset of this book project, the real questions were: (i) is this supply mix (i.e., the BAU scenario), good enough in terms of desired CO_2 reductions by 20,230, and (ii) is the current energy policy of Pakistan in tune to achieve a low-carbon economy in Pakistan. Based on our analysis, we did not see the answers in the affirmative. Then, it was natural to explore the possible pathways for Pakistan to achieve these objectives. In the previous sections, we evaluated both the alternatives (i.e., ILC and NPD scenarios) in terms of their potential for CO_2 reductions and affordability. Table 7.10 list the amount of CO_2 emissions in each of the scenarios for the period from 2020 to 2035. Among the three, the NPD scenario (i..e, the scenario with a dominant share of nuclear power (i.e., 32%) in the electricity supply mix) performs the best—not only meets the PA-based CO_2 reduction targets but also generate even lower electricity-related CO_2 emissions by a margin of 21% in the compliance year, 2030. For the period from the year 2030 to the year 2035, the NPD scenario generates 50% fewer CO_2 emissions than the renewable-dominant, ILC, scenario. Under this scenario, there appears to be a possibility of a low-carbon economic regime in Pakistan.

When it comes to the reliable supply of electricity, while nuclear power plants generate electricity 24/7 in the long-term (with low maintenance-based shutdowns), the renewable power sources (i.e., hydro, solar, and wind technologies), due to their intermittent nature and heavy dependence on weather make them a less reliable

Table 7.10 Electricity-related CO_2 emissions contribution by all scenarios

Year	Power-related emissions ($MtCO_2$)				Target[a] deviations ($MtCO_2$)
Scenario	2020	2025	**2030**[a]	2035	
BAU	112.8	517.5	**409.9**	2095.4	**+ 347.3**
ILC	113.3	367.6	**50.4**	35.7	**−13.9**
NPD	112.9	365.2	**43.3**	17.3	**−21**

[a]Target = 64.3 $MtCO_2$ by 2030

supply. With the ongoing developments and innovations in various storage technologies, the reliability factor for these renewable-based generation technologies will certainly improve. Therefore, once again, on the reliable supply of electricity front, ***the NPD scenario stands out*** clearly.

Pakistan is a developing nation. Like other developing countries, affordability is based upon—how much is the price of electricity, which plays a fundamental role in the demand for electricity both by people and enterprises. Figure 7.14 display the dynamic of the electricity prices in all the scenarios. In the post-2021 era, the existing Energy Policy of Pakistan, the BAU scenario, produced the most expensive electricity (e.g., due to prevailing circular debt, rising capacity charges, and higher fuel-import bills). Between the two alternatives, the NPD scenario generates electricity at a relatively higher price than the ILC scenario. Although the difference in prices lessens substantially when more nuclear capacity is operational (i.e., in the post-2026 era). However, a simultaneous look at Table 7.10 and Fig. 7.14 does favor the NPD scenario as the possible pathway for the policymaker of Pakistan.

Overall, with a convincing edge over **reliable** and **cleaner** electricity generation aspects and a **competitive price** regime, nuclear power turns out to be the technology of choice for policymakers in Pakistan. Given the dominance of thermal generation in India and KSA, it is not unreasonable to suggest aggressive investments in nuclear power technology for both countries. In the pursuit of nuclear power, a collaborative approach among the PIS region countries, therefore, can lead to a peaceful and greener region.

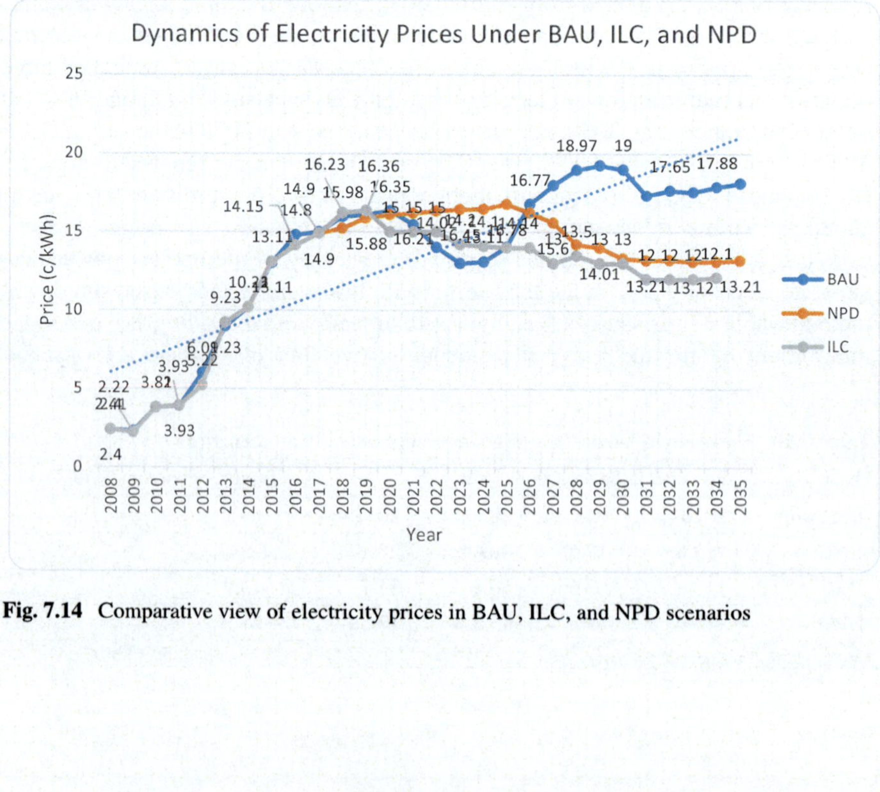

Fig. 7.14 Comparative view of electricity prices in BAU, ILC, and NPD scenarios

7.7 Summary of Chapter 7: The Key Insights

Climate change and global warming are severely affecting the well-being of people and economies across the globe. Strategies and actions required to address climate change issues including concerted efforts by countries to reduce CO_2 emissions as were agreed upon, voluntarily though, in the Paris Agreement. Most of the countries submitted their INDCs[13] with stipulated targets for the reduction of CO_2 emissions. In addition, countries are planning and striving for achieving low-carbon economies. In this context, here we evaluated the current Energy Policy of Pakistan and propose some viable alternative course of action to meet reduction targets for electricity-related CO_2 emissions and move towards a low-carbon economic regime in Pakistan. Key insights are:

- Methodologically, the dynamic modeling-based (e.g., MDESRAP in our case) scenario approach is a powerful decision-aiding tool for the systematic, socio-economic and environmental, and system-wide assessment and design of the energy policy.
- For the PIS region countries (i.e., Pakistan, India, and Saudi Arabia), the reduction of CO_2 emissions is an "*easier said than done*" case.
- The PIS region countries, all have ambitious plans for nuclear power expansion.
- Thermal power generation has the major share in the electricity supply mix of Pakistan, India, and Saudi Arabia.
- The current Energy Policy of Pakistan is not likely to meet its pledged reduction of CO_2 reductions as is the case of India and Saudi Arabia.
- Under the current energy policy incentives, thermal generation gains the major share (over 61%) of the electricity supply mix of Pakistan by the year 2030.
- Under the BAU scenario, the financial incentives alone (e.g., power purchase agreements with IPPs) are not enough to spurt the addition of clean and renewable electricity generation technologies in the overall electricity supply mix of Pakistan.
- In Pakistan's case, instead of meeting the target reduction of CO_2 emissions, its electricity sector will produce 347.3 $MtCO_2$ more (or above) than the target, as agreed in its INDC.
- Without a reasonable re-adjustment of the current policy incentive and initiatives, the policymakers in Pakistan most likely will not have the stated objectives of supplying reliable, affordable, and cleaner electricity. Instead, under the current conditions, Pakistanis will continue to pay higher electricity, face planned and unplanned blackouts and outages, and have more electricity-related CO_2 emissions.
- The locally renewable resource-intensive, ILC scenario appears to enable Pakistan to meet the target reduction in CO_2 and generates 13.9 $MtCO_2$ less than the target.

[13] INDCs: Intended Nationally Determined Commitments.

- The nuclear power dominant, NPD scenario, can generate the cleanest electricity; with over 94% share of clean and renewable generation in the electricity supply mix of Pakistan.
- The implementation of the planned carbon tax in Pakistan can make, otherwise capital-intensive, nuclear technology prices competitive with the renewable generation of electricity.
- While both alternatives, the ILC and the NPD scenarios produce relatively cheaper electricity than the BAU scenario (i.e., the existing Energy Policy of Pakistan), the cheapest electricity is generated under the ILC scenario.
- The NPD scenario can generate the most reliable, affordable, and cleaner electricity in Pakistan and can put Pakistan on track to achieve a low-carbon economy afterward of the 2030 era.

References

Brown, T. W., Bischof-Niemz, T., Blok, K., Breyer, C., Lund, H., & Mathiesen, B. V. (2018). Response to "Burden of proof: A comprehensive review of the feasibility of 100% renewable-electricity systems." *Renewable and Sustainable Energy Reviews, 92*, 834–847.

CAT. (2021). *Climate action tracker (CAT): Emissions in Saudi Arabia*. https://climateactiontracker.org/countries/saudi-arabia/. Accessed 23 Mar 2021.

Dyner, I., Franco, J., & Cardenas, M. (2013). Modeling progress towards emissions mitigation: Modeling low-carbon power generation policy. In H. Qudrat-Ullah (Ed.), *Energy policy modeling in the 21st century*. Springer.

EPA. (2021). *The United States environmental protection agency*. Greenhouse Gas Equivalencies Calculator. http://www.epa.gov/cleanenergy/energy-and-you/affect/oil.html. Accessed 29 Dec 2021.

Fatima, U., & Nasim, A. (2019). *Cost of electricity generation in Pakistan—Comparison of levelized cost of electricity of CPEC coal plants with oil and natural gas based plants commissioned in 2010–14* (Working Paper No. 02-19). https://ideaspak.org/wp-content/files_mf/1569051386IDEAS_WP2_Final.pdf. Accessed 17 Oct 2021.

Hansen, K., Breyer, C., & Lund, H. (2019). Status and perspectives on 100% renewable energy systems. *Energy, 175*, 471–480.

IEA. (2021). *India energy outlook 2021*. World Energy Special Outlook Report. https://www.iea.org/reports/india-energy-outlook-2021. Accessed 13 Oct 2021.

IPCC. (2008). *Intergovernmental panel on climate change*. http://www.ipcc-data.org/ddc_definitions.html. Accessed 13 Dec 2021.

Jahangir, M., Mostafaeipour, A., Habib, H., Saghaei, H., & Waqar, A. (2021). Effect of emission penalty and annual interest rate on cogeneration of electricity, heat, and hydrogen in Karachi: 3E assessment and sensitivity analysis. *Journal of Engineering*, Article ID 6679358, 1–16.

Mahmood, N., Wang, Z., & Zhang, B. (2020). The role of nuclear energy in the correction of environmental pollution: Evidence from Pakistan. *Nuclear Engineering and Technology, 52*(6), 1327–1333.

Mahmoud, M., Liu, Y., Hartmann, H., & White, D. (2009). A formal framework for scenario development in support of environmental decision-making. *Environmental Modelling and Software, 24*(7), 798–808.

Means, E., Patrick, R., Ospina, L., & West, N. (2005). Scenario planning: A tool to manage future water utility uncertainty. *Journal American Water Works Association, 97*(10), 68–75.

Mengal, A., Mirjat, H., Walasai, D., Khatri, A., Harijan, K., & Kaili, A. (2019). Modeling of future electricity generation and emissions assessment for Pakistan. *Processes, 7*(212), 1–26.

Mirjat, N. H., Uqaili, M. A., Harijan, K., Walasai, G. D., Mondal, M. A. H., & Sahin, H. (2018). Long-term electricity demand forecast and supply-side scenarios for Pakistan (2015–2050): A LEAP model application for policy analysis. *Energy, 165*, 512–526.

Nicholas, W., & Buckley, T. (2018). *The Institute for Energy Economics and Financial Analysis.* Pakistan's Power Future. http://ieefa.org/wp-content/uploads/2018/11/Pakistans-Power-Future_December-2018.pdf. Accessed 20 Mar 2021.

NTDC. (2018). *Electricity demand forecast based on multiple regression analysis.* National Transmission and Distribution Company, Government of Pakistan. http://www.mocc.gov.pk/SiteImage/Policy/National%20Climate%20Change%20Policy%20of%20Pakistan%20(2).pdf. Accessed 27 Nov 2020.

PNR. (2018). *Pakistan nuclear regulatory authority annual report 2018.* PNR Authority—Nuclear Regulatory Authority.

Qudrat-Ullah, H. (2015). Independent power (or pollution) producers? *Electricity Reforms and IPPs in Pakistan, Energy, 83*(1), 240–251.

Qudrat-Ullah, H. (2022). A review and analysis of renewable energy policies and CO_2 emissions of Pakistan. *Energy, 238*, 121849.

Rahman, M., Al-Ismail, F., Haque, M., Shafiullah, M., Islam, R., Chowdhury, M., Alam, M., Razzak, S., Ali, A., & Khan, Z. (2021). Electricity generation in Saudi Arabia: Tracing opportunities and challenges to reducing greenhouse gas emissions. *IEEE Access, 9*, 116163–116182.

Sadiqa, A., Gulagi, A., Bogdanov, D., Caldera, U., & Breyer, C. (2021). Renewable energy in Pakistan: Paving the way towards a fully renewables-based energy system across the power, heat, transport, and desalination sectors by 2050. *IET Renewable Power Generation, 16*(1), 177–197.

Timilsina, G. (2020). *Demystifying the costs of electricity generation technologies* (Policy Research Working Paper 9303). World Bank Group. https://openknowledge.worldbank.org/handle/10986/34018?locale-attribute=es. Accessed 19 Oct 2021.

Trainer, T. (2012). Can Australia run on renewable energy? The negative case. *Energy Policy, 50*, 306–314.

Trainer, T. (2013). Can Europe run on renewable energy? A negative case. *Energy Policy, 63*, 845–850.

WNA. (2014). *World Nuclear Association: Nuclear power in Pakistan.* http://www.world-nuclear.org/info/Country-Profiles/Countries-O-S/Pakistan/. Accessed 21 Apr 2021.

WNA. (2020). *World Nuclear Association: Nuclear power in the world today.* https://www.world-nuclear.org/information-library/current-and-future-generation/nuclear-power-in-the-world-today.aspx. Accessed 15 Jan 2021.

Chapter 8
Finale: Conclusions and Future Research Directions

8.1 Introduction

Whether one believes in climate change and global warming or not, the cariogenic, health-damaging, economically crippling (e., long droughts, more frequent floods, and extreme weathers) effects of environmental emissions are well known. Leaders and decision-makers of the world are striving to curtail the onslaught of CO_2 emissions. They agreed to the Paris Agreement (PA) in 2015. Most of the countries of the world, including all the three: Pakistan, India, and Saudi Arabia countries of the PIS region, are signatory to this agreement and have submitted their plans and initiatives for the reductions of CO_2 in their INDCs. Pakistan submitted its INDC[1] document in 2016. Therefore, as a genesis of this book, the overall objective of this book was set "to enhance systematically our understanding of and gain insights into, the general process by which the renewables and nuclear power's role in reducing CO_2 emissions can explain the making of a low carbon economy and a systematic achievement of PA's targets of a country." In our attempt to achieve this objective, we have presented, through Chapter 1 through Chapter 7, the systematic description of the nature and complexity of the issue of the reduction of electricity-related CO_2 emissions, a solid review of the role of modeling and simulation in the service of energy systems management, systematic formulation, and modeling of electricity supply systems, assessment of existing energy policy incentive and initiatives for the compliance of the pledged reduction of CO_2 emission, design, development and, and evaluation of reliable, affordable, and cleaner supply of electricity (i.e., with the detailed dynamics of nuclear power and renewable-based electricity supply alternatives) options for the energy domain decision makers and policymakers. To what extent we have succeeded in meeting the stated objective of this book, you, the reader, be the judge (feel free to provide me with your feedback at hassaq@yorku.ca). In this chapter, we present some major conclusions, key insights, and future research directions.

---

[1] INDCs: Intended Nationaly Determined Contributions.

H. Qudrat-Ullah, *Understanding the Dynamics of Nuclear Power and the Reduction of CO_2 Emissions*, https://doi.org/10.1007/978-3-031-04341-3_8

8.2 Major Conclusions and Key Insights

Understanding the dynamics of electricity supply, indigenous (and imported) resources (i.e., fuel), affordability (e.g., prices), and the environmental emissions (e.g., CO_2 emissions) is a complex dynamic task that energy policymakers of all the countries face. The case of Pakistan is no different. Therefore, consistent with the objectives of this book, (i) we took the challenge of assessing the current Energy Policy of Pakistan (ii) designing the viable alternate pathways for the best electricity supply mix of Pakistan that allow the policymakers to meet the pledged targets of CO_2 reduction and move toward a low-carbon economic regime, and (iii) assessing the impact of the increased share of nuclear power (i.e., at least 30%) on the electricity supply of Pakistan. We developed and validated a system dynamics-based dynamic model, MDESRAP, to carry out a systematic analysis to address this challenge.

Our dynamic model, MDESRAP, is suitable for making a range of electricity policy analyses including system-wide multidimensional assessment of any energy policy and design and development of alternate policies subject to the stated objectives and constraints. The integrated representation of the sectors for electricity demand, capital investments, production capital, fuel resources, electricity production, environment, and finance (costs and prices) in the model adequately allows for possible feedback effects both from within a sector as well as across the sectors. In addition, the effects of macroeconomic constraints, such as carbon tax, on the electricity supply sector and the environment are represented satisfactorily. The uncertainties, non-linearities, the dynamic interactions among the variables of an electricity system, and their potential effect on the electricity generation system are fully accounted for and represented in our dynamic model. With this structure, it becomes possible to assess the energy-economy-environment-wide implications as well as the costs and benefits of different electricity policy mix scenarios. *Therefore, the methodological conclusion of this study is that a system dynamics-based model can serve as an effective tool for electricity policy analysis.*

In addition, the modeling and simulation community, to enhance the utility and appeal of their simulation models, can utilize the validation procedures and tests that are demonstrated in Chapter 4 of this book. Novice learners of system dynamics methodology can benefit from Chapter 3 of this book where we have described in detail the model, MDESRAP, and its formulation along with the complete sets of underlying mathematical equations, sector by sector. Besides the development of this decision-aiding tool for the decision makers in the energy domain, here we provide our major conclusions and useful insights:

- With the existing energy policies and initiatives and the resulting electricity supply mix, *none of the PIS region countries: Pakistan, India, and Saudi Arabia are likely to meet its PA-based CO_2 reduction targets*. Without accounting for social, economic, environmental, and technical factors, any energy policy or initiative to reduce electricity-related CO_2 emissions is bound to fail. When we look at the electricity supply mix of any of these three countries under current conditions, the thermal (i.e., fossil-based) generation has the dominant share (e.g., over 60% is the share of coal-fired power plants in Pakistan. Therefore, *the energy policymakers of these countries, utilizing MDESRAP-based rigorous analysis, can begin the re-adjustment of their policy incentives to spur the cleaner and renewable generation of electricity as a substitute to their existing highly polluting thermal generation.*

- Based on our comparative analysis of the three options: (i) staying the course (i.e., no substantial adjustment or change in the existing energy policy invectives), (ii) indigenous clean and renewable resource-focussed alternate, and (iii) nuclear power dominant scenario, it appears that Pakistan is better off (i.e., in terms of reliable availability, affordability, and cleaner supply of electricity), with adding more nuclear power plants to its electricity generation capacity. With aspiring growth, plans for nuclear power growth, rising electricity demand, and with the stated objective of CO_2 reduction both India and Saudi Arabia can pursue the nuclear-power-dominant pathway for their future electricity supply mix.
- With a dominant share of nuclear power generation (i.e., a 32% share of nuclear power was produced in the nuclear power dominant scenario, as was discussed in our MDESRAP-based analysis of results which is reported in Chapter 7 of this book), Pakistan can not only meet its PA-based pledged reduction of CO_2 emissions but can embrace a path toward a low-carbon emitting electricity supply mix and a low-carbon economic regime.
- Under the current Energy Policy of Pakistan, the government's electricity purchase preferred rate, by itself, appears not to provide enough attraction and a stable signal for investments in renewable-based electricity generation. Instead, it has (e.g., in the pre-2020 period) already promoted risk-averse behavior on the part of the IPPs and private investors, who have predominantly invested in the relatively quick-return generating thermal power plants. As a result, neither electricity-related CO_2 emissions are decreasing in Pakistan nor the consumers of electricity in Pakistan are getting any relief at the electricity prices. Again, a clear call for the re-adjustments or a re-design of the Energy Policy of Pakistan.
- Although both the alternate policy scenarios: ILC (i.e., indigenous cleaner and renewable resources focussed) and NPD (i.e., the dominant share of nuclear power focussed) that are proposed and evaluated in this book can enable Pakistan to meet its CO_2 reduction targets and achieve a lower carbon economic regime, the aspect of having a reliable supply of electricity makes the NPD scenario as the preferred choice for the policymakers of Pakistan. Nuclear power plants are known for the reliable supply of electricity due to their comparatively less frequent maintenance and re-fuelling-based showdowns (Mahmood et al., 2020).
- On the affordability front, the BAU scenario produces much more expensive electricity as compared with the other two alternatives (e.g., in the ILC scenario, consumers can save about 24% on the electricity bill annually). Although the ILC scenario provides cheaper electricity than the NPD scenario, progressive implementation of planned *Carbon Tax* in Pakistan can make the NPD scenario equally price-competitive. Together with the availability of indigenous fuel and workforce ability, Pakistan is well set for adding more nuclear power to its electricity supply

mix. The notable benefits include a significant reduction in CO_2 emissions, a drastic decrease in the fuel imports dependency, and a relatively lower price and reliable supply of electricity for the consumers.

- Contrary to the end-of-the-pipe measures for CO_2 emissions reductions, the policy incentives (i.e., a rewarding premium for cleaner electricity generators and non-imported fuel users) that promote the competitive substitution for the least emitting technology may, in the long run, cause the economy to evolve with the reduced CO_2 emissions and the least imports dependency (i.e., due to fuel oil imports). Such re-adjustments in the existing policies and initiatives of any country are consistent with USDG number 7: availability of affordable and cleaner energy for all (Qudrat-Ullah, 2022). *Therefore, progressive investments in the least emitting technologies such as nuclear, hydro, solar, and wind technologies should be at the top of any, including the PIS region, government's electricity reform agenda.*
- Methodologically, we believe that our proposed and demonstrated combination of system dynamics modeling and multidimensional scenario-based approach can effectively contribute to the ongoing CO_2 reduction policy development efforts. Additionally, other areas of the energy policy domain that exhibit considerable complexity and conflicting interests such as energy conservation, energy storage, suitable supply of electricity across the regions, and regional energy infrastructure development planning, could benefit from this methodology.
- Although our MDESRAP-based analysis is based on Pakistan's case, the focus of this book, at times, we have also provided a comparative analysis of the situation in India and Saudi Arabia. Therefore, any learning and insights that this book presents, we hope it does ☺, are equally useful for the decision-maker and policymakers of India and Saudi Arabia. In fact, other jurisdictions, regions, and countries of the world with similar conditions (e.g., rising population, rising electricity demand, the dominant share of thermal generation, plans and aspiration to reduce CO_2 emissions and deploy cleaner and renewable electricity generating technologies) can benefit from the insights we have presented in this book to assess and re-design their energy policies to achieve reliable, affordable, and cleaner supply of electricity for the sustainable development and growth of their nations.

8.3 Future Research Directions

Energy decisions are highly context-dependent. The energy policy objectives of all countries and regions vary and depend on several factors other than just "meeting the demand"—the key objective of all energy policies. The other factors include (i) use of local resources, (ii) increasing employment ability, (iii) development of local capabilities, (iv) geopolitical considerations, (v) energy security, and (vi) bringing economic, financial, environmental, and social improvements across all the strata of their population. Future research for energy policy should account for these factors. It would be interesting to see how these factors will interact with "meeting the demand"

and "investors' perspectives" in the search for the best electricity supply mix of a nation.

When CO_2 tax is considered and an equivalent reward is given to clean energy production, nuclear power becomes price-competitive with solar and wind power technologies. As land use per KWh of electricity generated is least in the case of nuclear power, future studies can use appropriate costing of "land use" for each of the competing technologies to assess the long-term viability for clean energy technologies.

When stakeholders actively participate in the decision-making process, the sustainability and ownership of the resulting "solution" improves. Energy policy decisions should not be treated as an exception. We intend to investigate in future research projects: how does the active participation of the stakeholders in the energy decision-making process affect the implementation and sustainable operation of the "suggested" energy supply systems and solutions, over time.

Like all complex decisions, the decision-makers in the energy domain need decisional aids. System dynamics-based Interactive Learning Environments (ILEs) are in high demand (Qudrat-Ullah, 2015, 2022; Sterman, 2000). Simulation games are well known as an effective pedagogical tools at various levels of education. Our earlier research has shown the effectiveness of such ILES in improving and enhancing people's s decision making and learning in post-secondary education (Qudrat-Ullah, 2010). Our developed dynamic and validated model. MDESRAP may be used for the development of an ILE with electricity policymakers as its potential clients/users. For the education and training of our future leaders and decision makers in the energy domain, such decisional aids can also be integrated into the post-secondary curricula.

The development and promotion of renewable and cleaner production of electricity is a complex, dynamic task, which can be characterized by multiple stakeholders, actors, and conflicting objectives, values, and interests. In this book project, we have demonstrated the utility of a system dynamics model, MDESRAP, in assessing and designing energy policy. Future research can take the MDESRAP model[2] as a starting point and can explore its application to study the development and diffusion of various smart green technologies. The modular and generic structure of MDESRAP is flexible enough to be applied to the model assessment and design of energy policies for various jurisdictions and countries, which are focused on smart green technologies.

Although the focus in the design and development of MDSERAP was to evaluate the current electricity supply mix of Pakistan and design an alternate policy conducive to desired CO_2 reductions and in the long-term, move towards a low carbon economy, the electricity transmission system is also in dire need of investments for its renewal and update to match it with the expanded expected supply. Understanding the dynamics of transmission systems and associated investments critical for the sound design of an energy policy (Cincotti & Gallo, 2013) are complex Future researchers could develop a module related to electricity transmission capacity and investment

[2] As mdoels to be used in poliy making need not to replicate the various problem behaviors or patterns (Saeed, 2013).

dynamics. This added module will MDESRAP a suitable tool to carry out an integrated analysis of transmission and supply systems of the electricity sector of the country or region.

In our current dynamic model, MDSRAP, the import, and export of electricity are not explicitly considered. Future research can expand the boundary of MDESRAP to include these important aspects of energy policy assessment and design. Consequently, this proposed modification in the structure of MDERAP will enable its users to conduct a comprehensive assessment and design of energy policy for any nation or country.

Energy demand forecasting and estimation is an ongoing research activity. In our dynamic model. MSRP, we used a statistical approach for the endogenous generation of electricity demand. With recent advancements in the artificial intelligence and machine learning domain, future researchers could use, say, deep learning (Adewuyi et al., 2020) as a technique for the estimation of electricity demand.

References

Adewuyi, S., Aina, S., & Oluwaranti, A. (2020). A deep learning model for electricity demand forecasting based on tropical data. *Applied Computer Science, 16*(1), 5–1. https://doi.org/10.23743/acs-2020-01

Cincotti, S., & Gallo, G. (2013). Power plant relocation policy versus investments in transmission network infrastructure: A study on the Italian energy market. In H. Qudrat-Ullah (Ed.), *Energy policy modeling in the 21st century*. Springer.

Mahmood, N., Wang, Z., & Zhang, B. (2020). The role of nuclear energy in the correction of environmental pollution: Evidence from Pakistan. *Nuclear Engineering and Technology, 52*(6), 1327–1333.

Qudrat-Ullah, H. (2010). Perceptions of the effectiveness of system dynamics-based interactive learning environments: An empirical study. *Computers & Education, 55*, 1277–1286.

Qudrat-Ullah, H. (2015). Independent power (or pollution) producers? Electricity reforms and IPPs in Pakistan. *Energy, 83*(1), 240–251.

Qudrat-Ullah, H. (2022). A review and analysis of renewable energy policies and CO_2 emissions of Pakistan. *Energy, 238*(Part B), 121849.

Saeed, K. (2013). Managing the energy basket in the face of limits: A search for operational means to sustain energy supply and contain its environmental impact. In H. Qudrat-Ullah (Ed.), *Energy policy modeling in the 21st century*. Springer.

Sterman, J. (2000). *Business dynamics: Systems thinking and modeling for a complex world*. McGraw-Hill.

dynamics. This added module will MDESRAP a suitable tool to carry out an integrated analysis of transmission and supply systems of the electricity sector of the country or region.

In our current dynamic model, MDSRAP, the import, and export of electricity are not explicitly considered. Future research can expand the boundary of MDESRAP to include these important aspects of energy policy assessment and design. Consequently, this proposed modification in the structure of MDERAP will enable its users to conduct a comprehensive assessment and design of energy policy for any nation or country.

Energy demand forecasting and estimation is an ongoing research activity. In our dynamic model. MSRP, we used a statistical approach for the endogenous generation of electricity demand. With recent advancements in the artificial intelligence and machine learning domain, future researchers could use, say, deep learning (Adewuyi et al., 2020) as a technique for the estimation of electricity demand.

References

Adewuyi, S., Aina, S., & Oluwaranti, A. (2020). A deep learning model for electricity demand forecasting based on tropical data. *Applied Computer Science, 16*(1), 5–1. https://doi.org/10.23743/acs-2020-01

Cincotti, S., & Gallo, G. (2013). Power plant relocation policy versus investments in transmission network infrastructure: A study on the Italian energy market. In H. Qudrat-Ullah (Ed.), *Energy policy modeling in the 21st century*. Springer.

Mahmood, N., Wang, Z., & Zhang, B. (2020). The role of nuclear energy in the correction of environmental pollution: Evidence from Pakistan. *Nuclear Engineering and Technology, 52*(6), 1327–1333.

Qudrat-Ullah, H. (2010). Perceptions of the effectiveness of system dynamics-based interactive learning environments: An empirical study. *Computers & Education, 55*, 1277–1286.

Qudrat-Ullah, H. (2015). Independent power (or pollution) producers? Electricity reforms and IPPs in Pakistan. *Energy, 83*(1), 240–251.

Qudrat-Ullah, H. (2022). A review and analysis of renewable energy policies and CO_2 emissions of Pakistan. *Energy, 238*(Part B), 121849.

Saeed, K. (2013). Managing the energy basket in the face of limits: A search for operational means to sustain energy supply and contain its environmental impact. In H. Qudrat-Ullah (Ed.), *Energy policy modeling in the 21st century*. Springer.

Sterman, J. (2000). *Business dynamics: Systems thinking and modeling for a complex world*. McGraw-Hill.